EVERYDAY PRACTICE OF SCIENCE

EVERYDAY PRACTICE
OF SCIENCE

Where Intuition and Passion Meet
Objectivity and Logic

Frederick Grinnell

OXFORD

UNIVERSITY PRESS

2009

OXFORD
UNIVERSITY PRESS

Oxford University Press, Inc., publishes works that further
Oxford University's objective of excellence
in research, scholarship, and education.

Oxford   New York
Auckland   Cape Town   Dar es Salaam   Hong Kong   Karachi
Kuala Lumpur   Madrid   Melbourne   Mexico City   Nairobi
New Delhi   Shanghai   Taipei   Toronto

With offices in
Argentina   Austria   Brazil   Chile   Czech Republic   France   Greece
Guatemala   Hungary   Italy   Japan   Poland   Portugal   Singapore
South Korea   Switzerland   Thailand   Turkey   Ukraine   Vietnam

Published by Oxford University Press, Inc.
198 Madison Avenue, New York, New York 10016

www.oup.com

Oxford is a registered trademark of Oxford University Press.

Library of Congress Cataloging-in-Publication Data
Grinnell, Frederick.
Everyday practice of science : where intuition and passion meet objectivity
and logic / Frederick Grinnell.
p. cm.
Includes bibliographical references and index.
ISBN 978-0-19-506457-5
1. Science—Methodology.   2. Science—Social aspects.   I. Title.
Q175.G7538 2009
500—dc22   2008011104

1 3 5 7 9 8 6 4 2

Printed in the United States of America
on acid-free paper

In Loving Memory of
Anita Abraham Grinnell,
Nathaniel Grinnell &
Molly Cameron Azaroff

Science is serious play.

Leon Perkins, seventh-grade science teacher
Ardmore Junior High School, Ardmore,
Pennsylvania, 1956–1957

Preface

This book is based on the premise that the underlying nature of everyday practice of science should be understandable to anyone interested in science. Modern science education tends to ignore everyday practice and focuses instead on the linear model of scientific discovery—the *scientific method*. Rather than linear, the path to discovery in everyday practice tends to be highly ambiguous and convoluted. Real-life scientists begin their work situated within particular interests and commitments. Success requires intuition and passion as much as objectivity and logic. Scratch the surface of the anonymous and somewhat boring linear model and one finds excitement, risk, and adventure. In this book I describe everyday practice of science in a fashion that embraces intuition and passion without abandoning logic and objectivity.

What is the importance of understanding everyday practice of science? Popular magazines often have provocative front covers with headlines such as "How Medical Testing Has Turned Millions of Us into Human Guinea Pigs" (*Time*, April 22, 2002) or "Should a Fetus Have Rights? How Science Is Changing the Debate" (*Newsweek*, June 9, 2003). Advances in science and technology have turned questions previously part of religious and philosophical speculation into practical concerns of everyday life. As a result, one now finds contemporary science at the center of social and legal debates about diverse issues ranging from when life begins to when life ends. At the same time, incidents involving scientific misconduct and conflicts of interest have raised concerns about the integrity of scientists. In response to these developments, ordinary citizens, the mass media, and politicians want to know, "Are scientists doing research ethically?"

Many people accept the notion that the health, prosperity, and security of humankind depend on advances made possible by scientific research. Others are far less optimistic. They would argue that instead of further advances in science, changes in social and public policy will be necessary to have a positive impact on our common well-being. Moreover, the negative unanticipated consequences that sometimes accompany new scientific discoveries concern us all. Science brings about great benefits for humankind but also can lead to tragedy, such as the damage that technology has caused to our environment. Notwithstanding the potential for negative outcomes, an effective and innovative science and technology enterprise increasingly is viewed as prerequisite for maintaining national security, prosperity, and global competitiveness.

Two different perspectives inform my writing. I have an insider's knowledge of biomedical research based on more than 35 years of experience. My studies have been with subjects ranging from molecules to people. Over much the same period, I also have been studying philosophical issues at the interface of science, technology, and society, especially issues concerning biomedical ethics and research integrity. Combining these perspectives permits me to describe everyday practice of science and to integrate a philosophical and ethical dimension into my description. I frequently lecture on these subjects. My lectures form the basis for this book.

Science is not a monolithic activity, nor is there a single scientific community. What gets called science and how science is practiced have changed throughout history. In current issues of *Science* or *Nature*, articles can be found ranging from cosmology to economics. Biology research programs vary from highly descriptive to mathematical and theoretical. Despite the differences, practicing any kind of science requires value judgments: what to do, when to do it, how to do it, who should pay for it, and—after the work is completed—what the findings mean. I hope to provide insight into just these sorts of issues, even though most of the examples I discuss come from the biomedical sciences and the biomedical research community.

I organized *Everyday Practice of Science* into two parts: "Science" and "Science and Society." Each part has three chapters. Part I provides a description of practice: first an overview in chapter 1, and then a detailed account of the central activities of practice—discovery and credibility—in chapters 2 and 3. Discovery means learning new things about the world. Credibility means trying to convince others that the new findings are correct. Discovery and credibility circle around the researcher, whose biography and personality influence every step of these processes. Part II then turns to an analysis of several issues concerning science and society that in recent years have received significant national attention. Chapter 4 discusses integrity in science, from the societal level of science policy to the individual level of responsible conduct and conflict of interest. Chapter 5 focuses on informed consent and risk at the interface between human research and genetics. Finally, chapter 6 analyzes the relationship between science and religion. I suggest that science and religion represent distinct human attitudes toward experience based on different types of faith.

I wrote this book for a broad audience, including students, scholars, and the public interested in science. Individuals concerned about science education and science policy should find the work especially useful. Throughout the text, I avoid the use of jargon as much as possible. I include numerous citations to previously published material, more so than might be expected in a book written for a diverse audience. Rather than aiming for a detailed scholarly review of the recent literature in science studies, many of these references serve a more personal function. Most scientists I know complain about not having their discoveries appropriately cited by others. I often feel that way myself. Therefore, when I use ideas that I learned from others, I feel obliged to cite the relevant work if I still remember the original source. However, so as not to disrupt the flow of the chapters, these references are collected together at the end of the book.

I also include more quotations in the text than is common in a book about science. Many of the quotations that I use either come from the writings of Nobel laureates or concern Nobel Prize–winning research. I do not mean to imply that Nobel laureates are more knowledgeable about science than the overwhelming majority of researchers who have not achieved such recognition. I do not mean to imply that the Nobel Committee always is correct in its judgments about what scientists or scientific discoveries are most deserving of the Nobel Prize. Nevertheless, precisely because of their recognition, the Nobel laureates often have an opportunity to write memoirs about their lives in science. Nobel Prize–winning research often is the subject of analysis. I use these memoirs and analyses to illustrate key features of everyday practice of science. Their descriptions usually confirm my own experiences.

Acknowledgments

My thanks to Mark Frankel, Kenneth Pimple, William Snell, and Lee Zwanziger, who read and commented on the manuscript as the book evolved. Peter Farnham, Daniel Foster, David Hull, Stephen Jenkins, Mihai Nadin, Michael Ruse, Daniel Sarewitz, and Ken Yamada provided useful comments and suggestions. Charles Currin and the Southern Methodist University Ethics Colloquy offered important insights, as well. Former Oxford University Press editor William Curtis encouraged me to undertake this project. OUP editor Peter Prescott was of great help finalizing the work.

I am indebted to Richard Zaner, with whom I studied philosophy at Southern Methodist University in the 1970s, which was shortly after I joined the faculty at the University of Texas Southwestern Medical Center. Years later, Zaner arranged for me to study with his teacher Maurice Natanson when I was on sabbatical at Yale University. Together, Zaner and Natanson had a profound influence on my philosophical approach to understanding practice of science.

Finally, I thank Gayle Roth for her love and support during the final years of this project, and my children, Laura, Phillip, and Aviva, for their encouragement over the long time required to bring the book to completion.

Preparation of the manuscript was made possible by a grant from the National Library of Medicine (LM07526) and facilitated by a MERIT award from the National Institute of General Medical Sciences (GM031321). The ideas that I discuss are my own and do not represent the official views of the National Institutes of Health.

Contents

I
SCIENCE

1
PRACTICING SCIENCE

An Overview

Every year the U.S. National Science Foundation publishes a comprehensive analysis of *Science and Engineering Indicators*. As long as I can remember, the chapter on public attitudes contrasts two key points. First, Americans have a highly favorable opinion of science and technology. Second, Americans lack an understanding of basic scientific facts and concepts and are unfamiliar with the scientific process. Astronomer Carl Sagan called the situation "a clear prescription for disaster": "We live in a society exquisitely dependent on science and technology, in which hardly anyone knows anything about science and technology" (1).

In this chapter, I present an overview of the scientific process—what I call *everyday practice of science*. All of us practicing science face common problems: what to do, when to do it, how to do it, who should pay for it, and—after the work is completed—what the findings mean. I hope to provide some general insights throughout this book about these issues. Most of the examples I use come from biomedical research. If one wants "to piece together an account of what scientists actually do," wrote Nobel Laureate Sir Peter Medawar,

> then the testimony of biologists should be heard with specially close attention. Biologists work very close to the frontier between bewilderment and understanding. Biology is complex, messy and richly various, like real life.... It should therefore give us a specially direct and immediate insight into science in the making. (2)

I want to distinguish everyday practice from the idealized linear model of research. According to the linear model, the path from hypothesis to discovery follows a direct line guided by objectivity and logic. Facts about the world are there waiting to be observed and collected. The scientific method is used to make discoveries. Researchers are dispassionate and objective.

Although representative of the way that we teach science, I believe the linear model corresponds to a mythical account—or at least a significant distortion—of everyday practice. Rather than linear, the path to discovery in everyday practice is ambiguous and convoluted with lots of dead ends. Success requires converting those dead ends into new, exciting starts. Real-life researchers may aim to be dispassionate and objective, but they work within the context of particular life interests and commitments.

The two conversations of science

Figure 1.1 diagrams everyday practice of science. I place the individual scientist in the center. She engages in two conversations, one with the world to be studied, and the other with other members of the research community. The former conversation gives rise to the circle of discovery—learning new things. The latter gives rise

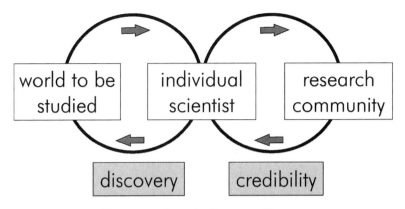

Figure 1.1. Everyday Practice of Science

to the circle of credibility—trying to convince others that the new findings are correct. These conversations are dialogs that proceed in an iterative manner. Of course, figure 1.1 is highly simplified because there are many conversations going on simultaneously. In addition, the researcher interacts only with a small part of the world, and the scientific community is itself within the world. Nevertheless, making the artificial distinctions in figure 1.1 helps to emphasize that there are important differences between these conversations. Interactions with the world typically are limited to making observations and carrying out experiments. Interactions within the research community depend largely on cooperative and competitive behavior.

Who is the individual scientist in figure 1.1? To help answer that question, I will introduce two imaginary researchers: Professor-It-Could-Be-Anybody and Professor-Somebody-In-Particular. Professor Anybody is the idealized researcher who does science according to the linear model. Professor Anybody is the scientist found in textbooks and research publications. Professor Particular, on the other hand, is the researcher engaged in everyday practice. Professor Particular experiences science as an adventure, so much so that she might write an autobiographical essay called "How to Get Paid for Having Fun" (3). A lot of us doing science feel just that way.

Science textbooks and research publications exclude everyday practice

There is no place for Professor Particular in the idealized structure of science. Sociologist Robert Merton described the norms of science as universalism, communism, disinterestedness, and organized skepticism (4). *Universalism* means that scientific claims are independent of the personal or social interests of researchers. *Communism* means that everyone owns scientific knowledge. *Disinterestedness* means that the community suppresses any tendency of investigators to behave according to their own self-interests. *Organized skepticism* means that researchers suspend and replace personal beliefs with an attitude oriented toward empirical and logical criteria. Merton's norms describe perfectly the characteristics of Professor Anybody: independent of personal or social interests, knowledge owned by everyone, disinterested, personal beliefs suspended.

It is Professor Anybody rather than Professor Particular who can be found in science textbooks. Textbooks usually present facts without clarifying where and how they arise. Space limitations may make this omission necessary. The consequence is that practice becomes invisible. The more common the knowledge, the more anonymous will be its source. Years of research are compressed into one or several sentences. At the same time, the adventure, excitement, and risks of real-life discovery disappear.

Research publications also mask the work of Professor Particular. To emphasize this point, I will describe some of the history of how researchers discovered messenger RNA. To understand this example, the following facts found in most modern biology textbooks will be useful:

- All cells store their genetic information in double-stranded molecules of deoxyribonucleic acid (DNA).

- Different cell types transcribe different portions of the sequence into specific messenger RNAs (mRNAs).

- These mRNAs then are processed and translated to make the proteins that determine in large part the specialized features of different cell types.

- Taken together, these steps represent the classic molecular information pathway of modern molecular genetics:

$$DNA \Rightarrow mRNA \Rightarrow protein$$

When a biology textbook states that mRNA is the intermediate between DNA and protein, the textbook sometimes adds a footnote to a 1961 research paper published in the prestigious scientific journal *Nature*. Evidence for the intermediary role of mRNA appeared first in the *Nature* paper (5). Research papers such as the publication in *Nature* provide the formal mechanism by which investigators report the details of their discovery claims to the scientific community.

The 1961 *Nature* paper about mRNA is titled "An Unstable Intermediate Carrying Information from Genes to Ribosomes for Protein Synthesis." The paper begins by summarizing prevailing views and controversies on the subject. Then the paper suggests a new hypothesis to resolve the controversial issues: "*A priori*, three types of hypothesis may be considered to account for the known facts." Experiments are proposed that could distinguish between the possibilities. Studies carried out are described. Conclusions are drawn from the results. The discovery claim is presented. Conceptually, the paper is arranged according to the sequence:

prevailing views ⇒ issues requiring further understanding ⇒ testable new hypothesis ⇒ experimental design ⇒ results ⇒ confirmation of one hypothesis and falsification of others

This sequence conforms to the linear model of science and gives rise to a paper whose plot will be none other than the scientific method. This plot is not, however, the way things actually happened. Rather, the scientific method represents a formal structure imposed upon what actually happened. "Writing a paper" wrote Nobel Laureate François Jacob—one of the authors of the *Nature* paper— "is to substitute order for the disorder and agitation that animate life in the laboratory.... To replace the real order of events and discoveries by what appears as the logical order, the one that should have been followed if the conclusions were known from the start" (6).

Stated otherwise, a research paper converts the process of discovery into an announcement of the discovery. In a sense, the paper itself becomes the discovery claim (7). Rather than discoverers, researchers become reporters of discoveries. They write "the (or these) data show" far more often than "our data show." Even if written in a personalized fashion, underlying every research report is the implication that that any scientist could have done the experiments and made the discovery.

Because the linear model of science typifies how scientists communicate with each other when they make public their research, the misimpression easily can arise that science actually proceeds in this fashion. Autobiographical writings of researchers provide a different perspective. In the case of the mRNA discovery, Jacob's view of what actually happened can be found in his memoir *The Statue Within*. Below are several quotes from Jacob's book followed by brief comments to highlight important features of everyday practice.

*We were to do very long, very arduous experiments. . . . But
nothing worked. We had tremendous technical problems.* (6)

In everyday practice, experiments can be divided into three classes:
heuristic, from which we learn something new; demonstrative,
which we publish—often repetition and refinement of heuristic
experiments; and failure, which includes Jacob's "nothing worked."
Not surprisingly, failed experiments represent the largest class.
Failed experiments arise for many reasons, including methodological
limitations, flawed design, and mistaken hypotheses. Experimental
failures are part of the normal process of science. Even inconclusive or uninterpretable results can still be extremely valuable if they
challenge the researcher's previous assumptions and teach her what
not to do the next time.

*Full of energy and excitement, sure of the correctness of our
hypothesis, we started our experiment over and over again.
Modifying it slightly. Changing some technical detail.* (6)

The objective and disinterested researcher envisioned by idealized
science would never be "sure of the correctness" of an unproven
hypotheses. Nevertheless, investigators' intuitions based on their previous knowledge and experience sometimes lead them to continue to
believe in and pursue a hypothesis even when the hypothesis appears
to be contradicted by the experimental results.

*Eyes glued to the Geiger counter, our throats tight, we tracked
each successive figure as it came to take its place in exactly
the order we had been expecting. And as the last sample was
counted, a double shout of joy shook the basement at Caltech.
Followed immediately by a wild double jig.* (6)

The exhilarating experience of success! Solving a challenging puzzle
and being the first to know the answer can elicit a degree of excitement

and enthusiasm uncharacteristic of serious grown men and women at work. When my seventh-grade science teacher, Mr. Perkins, told me "science is serious play," he was not exaggerating.

In summary, science comes in three different versions: (*i*) the facts—statements found in scientific textbooks—with little if any explanation of their source; (*ii*) the linear model—found in research publications and used by researchers to establish the credibility of their work and to influence the work of others; (*iii*) everyday practice—what really happened, a view rarely glimpsed by outsiders.

Science studies

In contrast to Merton's description of the idealized structure of science, philosopher Thomas Kuhn focused on individuals and their practices. Kuhn's book *Structure of Scientific Revolutions* had a great impact on development of the field called *science studies*. Rather than the idealized norms of science, the actual practices of individual scientific researchers and research teams became the focus of anthropologists, historians, philosophers, and sociologists, who together developed the field of science studies (e.g., 8, 9).

Kuhn described *paradigms* in science as sets of beliefs and values shared by members of a scientific community and as established and acceptable ways of problem solving (10). In addition, Kuhn emphasized that, beyond these criteria shared by the community, scientific judgment depends on individual biography and personality (11). Writers from other backgrounds also have emphasized the importance of individual biography and personality on how a researcher practices science. Examples include the *schemata* of psychologist Jean Piaget (12), *thought styles* described by physician-immunologist Ludwik Fleck (13), scientist-philosopher Michael Polanyi's *tacit knowledge* (14), and historian Gerald Holton's *thematic presuppositions* (15).

According to this way of thinking, the researcher's understanding of things is not simply given. Rather, understanding requires interpretation of experience. Interpretation takes place within the framework of one's life situation. Prior knowledge and interests influence what the person experiences, what she thinks the experiences mean, and the subsequent actions that she takes. Unlike idealized science, everyday practice can accommodate the remark that author Steve Martin has Einstein make to Picasso in Martin's play *Picasso at the Lapin Agile*: "What I just said is the fundamental end-all, final, not-subject-to-opinion absolute truth, depending on where you're standing" (16).

After *Structure of Scientific Revolutions*, science increasingly became of interest to study as an individual human activity characterized by, among other things, social and political aims. Given its potential impact on the world, understanding these aims would seem to be essential. Consider, for instance, questions that have been raised by the feminist movement (17). Upper-middle-class, white males have dominated science in the past and continue to do so in the present. Does this lack of gender diversity among researchers in the scientific workforce make a difference in the practice of science? If so, what difference? When it comes to getting a job, being promoted, or getting an equal salary, much evidence suggests that absence of role models and mentors has acted as a diversity barrier in science and engineering fields. Will lack of diversity also influence how science is practiced or what science is practiced?

Objectivity and the research community

Some people believe that the effect of cultural biases is limited to what is studied and not the conclusions reached by the research community. In his book *The Mismeasure of Man*, evolutionary biologist

Stephen Jay Gould argues otherwise. He uses historical examples from the late nineteenth and early twentieth centuries to describe how racist and sexist cultural attitudes influenced not only research design but also interpretation. Science progresses, wrote Gould,

> by hunch, vision, and intuition. Much of its change through time does not record a closer approach to absolute truth but the alteration of cultural contexts that influence it so strongly. Facts are not pure and unsullied bits of information; culture also influences what we see and how we see it. (18)

Gould's use of the expression *absolute truth* reflects the important distinction between truth (small "t") as we now understand things and Truth (capital "T") that no further experience will change. I will emphasize frequently that everyday practice of science is after truth. Science always is a work in progress, which makes the process exciting and challenging. Anyone who claims to know already the Truth of a matter must be depending on sources of information outside everyday practice of science.

An important example of cultural bias comes from the history of psychiatry. Until the early 1970s, homosexuality was viewed widely as an illness and was listed as such in the 1968 version of the *Diagnostic and Statistical Manual of Mental Disorders* (*DSM-II*) of the U.S. psychiatric community. That the diagnostic classification was political rather than scientific is shown by how the classification was changed. In 1973, the board of directors of the American Psychiatric Association *voted* that homosexuality was not an illness. Membership ratified that vote a few months later (19). With the link between homosexuality and psychopathology discredited, homosexual couples increasingly have been accorded the same rights and respect as heterosexual couples. Now those who oppose homosexuality, and many still do so, can less easily appeal to "scientific/medical facts" to support their objections.

For some, admitting the human associations of science can challenge the belief that science provides an objective description of reality. Especially at the fringe of the so-called postmodernist movement, the argument has been put forth that scientific facts are merely culture-dependent, normative beliefs. If there is truth to be learned, then scientific inquiry deserves no privileged status. Truth-for-the-individual likely is the best for which one may hope.

The postmodernists are wrong. Culture may influence what we see and how we see it, but the dramatic impact of technology on the world shows that much of scientific knowledge is more than mere belief. Throughout history, we humans have been attempting to overcome natural threats to our existence, such as famine and disease. Beginning with the discovery and use of fire and the invention of primitive tools, controlling and changing the environment has been a central human project. The ability of science to produce technologies with increasing impact on the world suggests that science's understanding of the physical mechanisms of the world has advanced.

So here is a paradox. How can practice of science situated within a particular cultural context give rise to knowledge that has universal validity? How does Professor Particular become Professor Anybody?

My way to begin to answer this question is by comparing scientific researchers with baseball umpires. According to tradition, there are three types of baseball umpires:

The first type says, "I call balls and strikes as they are."

The second says, "I call them as I see them."

The third says, "What I call them is what they become."

What distinguishes these umpires is not the situations in which they find themselves, but rather the attitudes that they bring to their work. Because of their different attitudes, they practice umpiring differently. The first emphasizes Truth; the second, context; the third, power.

Those who have learned the idealist, linear view of science frequently identify researchers with the first type of umpire. Postmodernists identify researchers with the third. Further description of discovery and credibility will clarify why the second type of umpire corresponds most closely to the way that scientists work.

In everyday practice, discovery begins in community. Community offers continuity with the past and interconnectedness of the present. Each researcher or group of researchers initiates work in the context of prevailing experiences and beliefs—the starting point and justification for further action. We assume that this previous knowledge is incomplete or to some degree incorrect. There is little reward in science for simply duplicating and confirming what others already have done. What we aim for is *new-search* rather than *re-search*.

What I am focusing on here is discovery at the frontier of knowledge, a place where no one has been before. At the frontier, one encounters an ambiguous world demanding risky choices. What should be done first? What is the difference between data and noise? How does one recognize something without knowing in advance how it looks? Of course, not all research occurs at the frontier. Clinical investigation involving humans should begin only in much more settled territory after a great deal of preclinical work has been accomplished. The ethics of research with humans demands that the work be as unambiguous as possible.

At the edge of knowledge, incomplete understanding can result in mistaken assumptions and errors in experimental design. At the same time, incomplete understanding sometimes permits observation of unexpected results. Nobel Laureate Max Delbrück called the latter aspect of research the *principle of limited sloppiness* (20). Here, sloppiness does not refer to technical error, although some important discoveries have their origins in just that fashion. Rather, Delbrück meant sloppiness in the sense that our conceptual understanding of a system under investigation is frequently a little muddy. Consequently, experimental design sometimes tests unplanned questions, as well as those explicitly thought to be under consideration. Unexpected results can emerge and lead to important findings if the experimenter notices (21). *We do more than we intend.* The underlying ambiguity of practice makes what we call luck or serendipity a frequent feature of discovery.

Because Professor Particular cannot avoid the possibility of error, including self-deception, her initial discoveries should be thought of as *protoscience.* For protoscience to become science, the researcher not only must be able to replicate her own work, but also must turn to the community to convince peers of the correctness of the new findings. Professor Particular overcomes her subjectivity through *intersubjectivity.* Intersubjectivity assumes *reciprocity of perspectives*—if you were standing where I am, then you would see (more or less) what I see. The world is ours, not mine alone (22).

Reciprocity of perspectives makes possible the process of credibility. Other researchers usually offer responses to discovery claims that can range from agreement to profound skepticism. They react to the specifics of the research as well as to the relationship between new ideas and prevailing beliefs. Novel and unexpected discovery claims sometimes will be rejected or unappreciated by the community because the new thinking does not fit current understanding. The history of Nobel Prize–winning research is replete with such examples.

Rather than accept rejection, to succeed in scientific research often requires that researchers become advocates for their work. When the awards are given out, we frequently read:

> Why were Professor Particular's early studies ignored, neglected, and often denigrated? ... The powerful force of the longstanding dogma made it easy for the community to brush

aside Particular's experiments and ideas and to view them as a curiosity with little or no relevance to the mainstream. Fortunately, Particular's passionate belief in his data and his unshakeable self-confidence propelled him forward despite the criticisms of his colleagues. (paraphrased from 23)

Of course, becoming an advocate for one's beliefs when everyone else thinks that you are mistaken is risky business. What appears to be novel often turns out to be experimental artifact. N-rays, polywater, and cold fusion bring to mind some of the most famous cases of erroneous research. The only thing worse than being wrong in science is being ignored. The former frequently leads to the latter.

In the end, Professor Particular becomes Professor Anybody through the process of credibility. During this process, investigators shape and reshape their work to anticipate and overcome the criticisms that they receive from the community (24). When (if) others eventually validate the new observations by using them successfully in their own research—often modifying them at the same time—then the new findings become more widely accepted. In short, credibility happens to discovery claims. Discovery claims become credible—are made credible or incredible—through their subsequent use (25).

Returning to the baseball umpire analogy, in everyday practice of science calling things as they are is reserved for the community rather than the individual. But even the community's calling is tentative. With discovery oriented toward completion and correction, the scientific attitude defers Truth to the future and aims for credibility in the present. The realism of science remains incipient and tightly linked to practice through last year's discoveries. Last year's discoveries become this year's conceptual and technological instruments of exploration. Thus, realism of science emerges not through power, as supposed by the postmodernist critique, but by replacing individual subjectivity with communal intersubjectivity, philosopher

Annette Baier's *commons of the mind*: "We reason together, challenge, revise, and complete each other's reasoning and each other's conceptions of reason" (26).

At the ideal limit, reciprocity of perspectives means that all scientists can share the same experiences. As experience becomes typical and commonplace, the unique individual disappears and the anonymous investigator (Professor Anybody) emerges. Scientific knowledge aims to be correct for anyone, anywhere, anytime.

In summary, objectivity of science does not depend on the individual. Rather, objectivity is a function of the community. Everyday practice of science is neither truth nor power, but rather balanced on a contextual ledge in between.

In practice, biography and personality never really disappear. Intersubjectivity can be achieved only partially. Because the objectivity of science depends on the community rather than the individual, the influence of personality and biography on the researcher's scientific judgments becomes an asset to science rather than an impediment. Diversity in how people think and work enhances scientific exploration of the world. Diversity of demographics—for example, gender, race, and economic status—enhances the possibility of a multicultural approach (27). Without diversity, the community cannot really "complete each other's reasoning and each other's conceptions of reason." The judgments of a research community that is too homogeneous or isolated are just as much at risk as those of a community prevented by political interference from open exchange and dissent.

The foregoing discussion emphasizes the inherent ambiguity of everyday practice of science. Table 1.1 explicitly contrasts this ambiguity with the stages of the classic scientific method. The ambiguity evident in table 1.1 highlights Medawar's comment that there is no such thing as the scientific method, and that the idea of naive or innocent observation is philosophers' make-believe (2). Inevitably, both sides of table 1.1 blend together.

Table 1.1. The Classic Scientific Method vs. the Ambiguity of Everyday Practice

The Classic View	The Ambiguous View
State the problem to be studied.	Choosing a problem commits one to investing time, energy, and money. The wrong choices can place one's life goals and career in science at risk.
Carry out experiments to study the problem and record the results.	The important results may not be noticed. What counts for data one day may appear to be experimental noise the next.
Conclude whether the observations confirm or falsify one's ideas.	If the results don't agree with expectations, it may be because the idea is wrong or because the method used to test the idea is flawed. Hence the adage: Don't give up a good idea just because the data don't fit.
Seek verification by other researchers of the findings and conclusions.	Discovery claims are often greeted with skepticism or disbelief, especially when they are very novel and unexpected. Rejection by other scientists is a common experience. To succeed, investigators frequently have to become advocates for their work.

Education without practice

We frequently hear the question, "What ails U.S. science and mathematics?" For more than a generation, an emphasis on the shortcomings and need for enhanced science education in the United States has been recognized in every national report that addresses the subject. The huge literature that has developed offers many answers to the foregoing question but lacks consensus. "The candidates include teachers who don't know the subject matter, lousy textbooks, a badly designed curriculum, low expectations by educators and parents, an outmoded school calendar, and the debilitating effects of poverty and race" (28). In addition, maybe students are just "turned off." They think of science as a mere collection of facts rather than as high adventure. "Dry as dust," commented Nobel Laureate Leon Lederman (29).

Shortly before his death in 1994, I heard Nobel Laureate Linus Pauling lecture at a science education workshop. Pauling began his personal reflection by holding up a contemporary college chemistry text. He suggested that the book was too thick—several inches too thick. In his view, textbooks had become collections of facts divorced from understanding.

Divorced from understanding reflects at least in part the omission of everyday practice from science education. This criticism is nothing new. More than 50 years ago, Harvard University President James Conant pointed out the problem in *Science and Common Sense*:

> The stumbling way in which even the ablest of the scientists of every generation have had to fight through thickets of erroneous observations, misleading generalizations, inadequate formulations, and unconscious prejudice is rarely appreciated by those who obtain their scientific knowledge from textbooks. (30)

Even the science fair, one of the most popular and valuable science education experiences, distorts practice. The science fair judge begins by asking, "Is the problem stated clearly and unambiguously?" The hypothesis always goes near the upper left-hand corner of the poster board describing the science project, and must come first—never last. When I encouraged one of my children to put the hypothesis at the lower right as her conclusion, she lost points. After that, she questioned whether I really understood science! Traditional science fairs reward success in research and clarity of presentation.

What kind of science fair rewards success in the playfulness of discovery, including learning what not to do the next time?

Why has everyday practice not become a more central focus for science education? Whatever the reasons, ignoring practice impedes the goals of science education. When he was executive director of the National Science Teachers Association, Bill Aldridge wrote that the framework for science education should be built around three fundamental questions: What do we mean? How do we know? Why do we believe? (31). Those who do not understand the practice of science cannot, in the end, answer these questions.

2

DISCOVERY

Learning New Things about the World

When I lecture about discovery in our university course on research integrity, I begin by reminding the audience that discovery is hard to accomplish, failure is frequent, and the pressure to produce is great. Everyone who is familiar with the expression "publish or perish" understands the third point. Science is very competitive. That "discovery is hard to accomplish" and "failure is frequent" are features less well appreciated. For the graduate students in the lecture hall, understanding these latter two features helps them deal with their own situations. Many of these students have been struggling to discover something new. Progress has been slow. They think everyone else is doing so much better. In the scientific papers they read, rarely will they find experimental failures described. They do not yet have sufficient experience to understand why so many experiments are inconclusive or uninterpretable—why 10 research notebooks' worth of experiments might be required to publish a 10-page research paper.

The word *discovery* has multiple meanings. We talk about discovering everything from fundamental laws of nature to rock stars. Public announcement of a discovery creates ownership for the discoverer. In the case of gold mines or patent applications, the discoverer gets legal and monetary credit. Legal and monetary benefits can be transferred to others. Something additional happens with scientific discoveries. The researcher or research team gets intellectual credit. Intellectual credit increases personal influence, power, and fame. Intellectual credit cannot be transferred and belongs only to the discoverers. In the tournament model of science, there can be only one winner (or one winning team) (1).

Of course, the question of who should get intellectual credit for a discovery sometimes becomes very controversial. In my field, there was dismay among many researchers that cell biologist Keith Porter did not share the 1974 Nobel Prize in Physiology or Medicine (2). That year's Nobel Prize recognized discoveries concerning the structural and functional organization of the cell, discoveries in which Porter played a key role. Above all else, the scientists I know want intellectual credit. They are very unhappy if they feel that their intellectual contributions have been ignored by others. Some of us respond by sending the perceived offenders copies of our own research papers: "Dear Professor: After reading your recent publication, I thought you would find interesting the attached work." Of course, not everyone responds directly. Some simply shrug off these occasions—"That's science." My former cell biology colleague and mentor Wayne Streilein would calm down the younger members of our department by saying, "You do the best you can in an imperfect world!"

The emotional thrill that accompanies discovery in science comes above all from the feeling that one has solved a challenging puzzle and is the first to know something new about the world. The intensity of that experience ranges from the commonplace to what sometimes are called "eureka" moments. Eureka differs from "ah-ha." Ah-ha lacks the intensity that I have in mind. Ah-ha comes after listening to the first few notes of a song and then suddenly recognizing the tune—"ah-ha, nice!" Eureka, by contrast, comes after reading a long and complicated novel in which loose strings hang out everywhere and confusion reigns until a single piece of information allows everything before to be reconstructed into a coherent framework—"brilliant!"

Even if the problem at hand seems very commonplace, exhibiting minimal ambiguity regarding potential outcomes, surprises sometimes happen. For example, laboratory research and animal studies suggested that the test chemical UK-92,480 might make a useful anti-angina medication. After preclinical work identifies a test chemical as a potential drug, the next step of clinical research is phase I safety trials. In phase I trials, human volunteers receive different doses of the experimental drug to test for risky side effects. Men who participated in the UK-92,480 safety trials reported that they developed penile erections! UK-92,480 became Pfizer's blockbuster drug Viagra, introduced to the public in 1998 (3).

The process leading up to the moment of discovery can be analyzed at multiple levels. These levels range from the description of a single experiment to a historical treatise. Figure 2.1 illustrates just such an example for the development of open-heart surgery (4). Open-heart surgery was attempted first in the 1950s after development of the heart-lung machine permitted patients to receive artificial life support. The successful cardiac surgeon stands at the peak of a mountain of previous discoveries. Each of these discoveries, including Benjamin Franklin's work on electricity (hence the title of the article, "Ben Franklin and Open Heart Surgery"), contributed to the ultimate success of the procedure. Figure 2.1 illustrates Sir Isaac Newton's famous comment to fellow scientist Robert Hooke, "If I have seen further, it is by standing upon the shoulders of giants."

Figure 2.1. Development of Open-Heart Surgery
From Comroe & Dripps, 1974. Ben Franklin and Open Heart
Surgery. *Circ Res.* 35. Reproduced with
Permission from Lippincott Williams & Wilkins.

The complexity of figure 2.1 should make clear that designating a particular moment as the moment of discovery is highly arbitrary. A partial analysis of the important work leading up to the development of the electrocardiogram—itself a necessary step for open-heart surgery—lists 45 papers, including the description in 1660 of the first electricity machine (5).

Throughout their careers, investigators continually make discoveries. They formalize these claims and present them to the larger scientific community when they believe that sufficient new information has been accumulated to support the claims. Making public a discovery claim is a key moment in the process of discovery (6). Going public creates intellectual ownership for the discoverers. Going public also gives up intellectual control. Once other researchers know about the work, they can use the new findings in their own investigations. The discovery often will be transformed into something quite different from what the original discoverers anticipated.

Scientific researcher as world explorer

Metaphors offer important insights into the nature of discovery. The metaphor of scientific researcher as world explorer is one of my favorite, especially this description of Christopher Columbus: "Columbus did not know where he was going when he started out, did not know where he was when he got there, and did not know where he'd been when he got back" (7).

Of course, Columbus thought he knew where he was going, where he had arrived, and where he had been. Based on his prior beliefs and expectations, he was able to convince the Spanish government to fund his project. Only afterward—as he and others attempted to reproduce the voyages—did it become evident that the original assumptions and conclusions were flawed. In science, initial misconceptions often provide the impetus for research that leads to discoveries different from those that the investigator thought he was looking for at the beginning of exploration.

The travels of *The Three Princes of Serendip* emphasize another important aspect of world exploration. According to this fairytale, the three princes' father sent them on a mission. During their travels, they encountered many things for which they were not searching.

Eighteenth-century writer Horace Walpole coined the word *serendipity* based on the fable (8). Likewise, unintended but potentially valuable encounters often occur during a scientist's experimental travels. Historical accounts of discovery frequently attribute an important role to serendipity. The introduction to *Serendipity: Accidental Discoveries in Science* begins: "What do Velcro, penicillin, X-rays, Teflon, dynamite, and Dead Sea Scrolls have in common? Serendipity! These diverse things were discovered by accident, as were hundreds of other things that make everyday living more convenient, pleasant, healthy, or interesting" (9).

Another metaphor—discovery at the *frontier of knowledge* (10)—emphasizes the image of the American West in the mid-nineteenth century. Life on the frontier was risky and uncertain. Survival depended on innovation, creativity, and flexibility. At the frontiers of research, even the distinction between data and noise sometimes will be unclear. Nonspecific background noise inherent in research methodology can be almost as large as the specific experimental result itself. To decide what counts as data, investigators frequently have to rely on intuition as much as on logic.

At the frontier, one usually can find the most excitement and greatest potential for experiencing the eureka of discovery. Historian Gerald Holton wrote that to understand the "exuberant enthusiasm" of research,

> one should go not to a well-established contemporary physical science, but perhaps to a field when it was still young. In the journals of the seventeenth and eighteenth centuries, we can find...a marvelous and colorful efflorescence of interests and an unselfconscious exuberance that verges sometimes on aimless play. The scientists of the time seem to us to have run from one astonishing and delightful discovery to the next, like happy children surrounded by gifts. (11)

The description of scientists as "happy children surrounded by gifts" would have appealed to Dan Koshland, former *Science* magazine editor. Koshland's imaginary editorial spokesperson, Dr. Noitall, once commented, "Scientists have failed to grow up. They are all children, eternally curious, eternally trying to find out how the pieces of the puzzle fit together, eternally asking Why, and then irritatingly asking Why again when they get the answer to the first question" (12).

Plato: Is discovery possible?

For those who think about scientific discovery in terms of logic and objectivity, my emphasis on ambiguity and serendipity may come as a surprise. Perhaps even more surprising will be learning of the philosophical challenge to the possibility of discovery. Examination of this challenge will make it clear why novelty and innovation are so difficult to achieve.

More than 2,000 years ago, the Greek philosopher Plato argued against the conventional idea of discovery. In the *Dialogues*, Plato has Meno ask Socrates:

> How will you look for it, Socrates, when you do not know at all what it is? How will you aim to search for something you do not know at all? If you should meet with it, how will you know that this is the thing that you did not know?
>
> [Socrates answers:] I know what you want to say, Meno... that a man cannot search either for what he knows or for what he does not know. He cannot search for what he knows— since he knows it, there is no need to search—nor for what he does not know, for he does not know what to look for. (13)

According to the interchange between Meno and Socrates, what we usually mean by discovery is impossible. If a person seeks and finds what he already knows how to look for and recognize, then nothing new has been discovered. On the other hand, he cannot find what he does not know how to recognize. Every scientist faces Plato's paradox. Inherent in everyday practice is the tendency for the researcher to discover precisely that for which he is looking.

The problem of noticing

Plato treated discovery as an event, not a process. As an event, discovery has two extremes: "it's one of them" or "I did not notice anything, where?" As a process, additional possibilities occur in between:

- I don't know what it is. I've never seen one before.
- It looks like one of them, but I'm not sure.

In everyday practice, discovery often begins when something is noticed but unrecognized or when something previously known is found in an unexpected place. Being prepared to notice the unexpected often is the key.

The famous (but fictional) detective Sherlock Holmes notices everything. His ability to notice everything is what makes him so appealing. The rest of us, like Holmes's partner, Watson, tend to overlook the unexpected or sometimes the absence of the expected, as occurred in *Silver Blaze*:

> Colonel Ross still wore an expression which showed the poor opinion which he had formed of my companion's ability, but I saw by the inspector's face that his attention had been keenly aroused....
>
> "Is there any point to which you would wish to draw my attention?"
>
> "To the curious incident of the dog in the night-time."

"The dog did nothing in the night-time."

"That was the curious incident," remarked Sherlock Holmes. (14)

After discovery finally does happen, one often hears, "Of course. How did I miss the answer for so long?" Nothing noticed—nothing discovered.

Figure 2.2 illustrates the problem of noticing. The figure shows a diagram of the complex apparatus used by Nobel Laureate Melvin Calvin to make quantitative and time-dependent measurements of the process of photosynthesis (15). One can find inserted in the diagram an out-of-place and comical element. Here is an opportunity

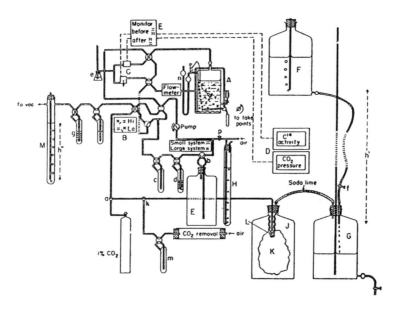

Figure 2.2. Melvin Calvin's Photosynthesis Apparatus
From Wilson & Calvin, 1955. The Photosynthetic Cycle. *J Am Chem Soc.* 77.
Reproduced with Permission from the American Chemical Society.

for the reader to try to notice the unexpected. When I show this image during a lecture, usually no more than 1 in 100 participants will be able to notice the inserted element in the few minutes that I project the image and discuss the problem of noticing.

In the case of figure 2.2, the inserted element went unnoticed by reviewers and journal production staff and remained in the version of the paper that was published in 1955 in *Journal of the American Chemical Association*. Even after publication, no one noticed for many years. Later in this chapter, I will reveal the element. Once known, the reader will never be able to see figure 2.2 without noticing the comical element, so enjoy looking at the figure in an unlimited fashion while you can.

Unintended experiments

Scientist and philosopher Michael Polanyi labeled the *tacit dimension* of knowledge the ability to learn something new without being explicitly aware of having done so—*we know more than we can tell* (16). Sometimes discovery simply requires "getting in touch with" that tacit knowledge. Nobel Laureate Albert Szent-Györgyi described this as seeing what everybody else has seen and thinking what nobody else has thought.

Getting the opportunity to notice something unexpected requires more. This opportunity often arises in everyday practice because we *do more than we intend*. Nobel Laureate Max Delbrück called this feature of science the *principle of limited sloppiness* (17). The reader should not be fooled into thinking that by sloppy Delbrück meant careless or imprecise. Rather, he had in mind the muddiness of our conceptual knowledge about the system under investigation. Experimental design frequently tests unintended questions as well as those questions explicitly under consideration. Unexpected results can emerge and lead to important findings if the experimenter notices (18).

An unintended experiment totally changed the direction of my own research. I had become a postdoctoral fellow in biochemist Paul Srere's laboratory at the Veteran's Administration Hospital in Dallas, Texas. Srere studied regulation of metabolic enzymes, particularly the enzymes of the citric acid cycle. The citric acid cycle provides the central mechanism of energy production for many types of cells. Like many biochemists, Srere studied enzyme regulation using isolated, purified enzyme preparations. Nevertheless, he was convinced that understanding regulation depended on studying enzymes in the context of their organization within cells.

My project was to use rat liver cells growing in cell culture to analyze changes in the cellular levels of citric acid cycle enzymes that occurred in response to altered energy metabolism. To alter energy metabolism, I planned to incubate the liver cells in nutrient-modified culture medium. In addition, I planned to add chemicals to the culture medium to modulate energy metabolism.

Initially, I carried out experiments to learn if any of the conditions that I wanted to use would either destroy the cells or prevent them from sticking to the surface of the cell culture dishes. Either of the latter consequences would make the treatments unacceptable for my purposes. Figure 2.3 shows a page from my notebook that describes the first preliminary experiment. The top half of the notebook page lists 12 different conditions that were tested, 11 on the upper row and 1 below. The bottom half of the page summarizes the results. Everything looks messy. I was taught to write down the planned experiment first and then to show in my notebook any changes that were made as the experiment progressed.

Line 10 toward the bottom of the page shows the results of treating cells with a chemical (arsenite) known to block energy metabolism. Under these conditions, none of the cells appeared to stick to

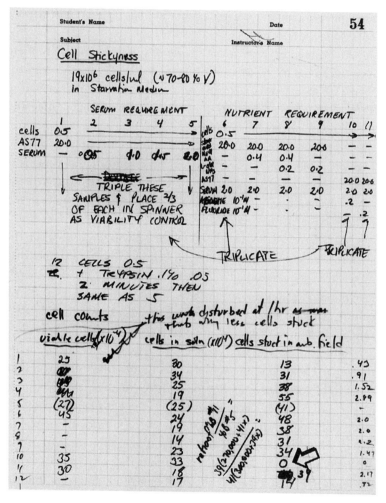

Figure 2.3. Preliminary Cell Sticking Experiment
From the Author's Notebook TC-I, p. 54, October 1970.

the dishes (an arrow points to the result). This is precisely the type of outcome about which I was concerned. Since the cells did not stick after treatment with arsenite, I would be unable to analyze enzyme levels under these conditions.

Instead of continuing with my originally planned experiments minus arsenite, I focused my attention on the difference between line 10 and line 11. I had expected these two conditions to have the same consequence for the cells since both involved treatment with chemicals that blocked energy metabolism. Why were the outcomes different? Other than not sticking, the cells did not appear to be damaged. I wondered if arsenite might be acting as a specific inhibitor of cell sticking. By entertaining this notion, I suddenly had switched research projects.

Experimental results can be thought of as pieces of a jigsaw puzzle. The investigator has in mind a particular puzzle when he begins an experiment, and the result usually fits into that puzzle. Nevertheless, every experimental result/puzzle piece can fit into more than one puzzle. In the Viagra example mentioned earlier:

Puzzle 1: Is the potential anti-angina drug UK-92,480 safe for use in humans?

Unexpected observation: Men taking UK-92,480 developed penile erections.

Puzzle 2: Is UK-92,480 a useful drug for treating penile dysfunction?

And restating the example from my research:

Puzzle 1: Does arsenite alter cellular levels of citric acid cycle enzymes?

Unexpected observation: Arsenite inhibits cell sticking.

Puzzle 2: Does arsenite block the molecular mechanism of cell sticking?

If the results are seen in retrospect as part of a different puzzle than that originally planned, then the investigator has carried out an unintended experiment. Once reconceptualized, the unexpected observation can be seen to answer a prediction or question based on the second puzzle instead of the first. If one suspected that an arsenite-sensitive enzyme played a role in cell sticking, then testing the effects of arsenite on cell sticking would be a direct way to test this idea.

The philosopher Charles Peirce recognized the significance of unintended experiments for scientific discovery and called this way of thinking *abduction*, as opposed to induction or deduction. According to Peirce's formal description of abduction (paraphrased from 19):

An unexpected fact C is noticed.

But C would be expected if A were the case.

Consequently, there are grounds to investigate A.

Abduction acts as a powerful general mechanism by which investigators gain new perspectives into their research. The consequences are emotional as well as intellectual. Being surprised is fun. Unintended experiments push the research in new directions, but precisely where the investigation will lead remains to be seen. Now we are indeed exploring the unknown. "The game is afoot, Watson!" as Holmes would say.

The problem at hand

Noticing a new potential problem for investigation does not necessarily lead a researcher to refocus his work. For Srere to let me switch my focus from enzyme regulation to cell adhesion was a risky decision on his part. Understanding the risk requires a discussion of the "problem at hand."

In deciding to investigate a problem, a researcher's previous knowledge and experience are critical factors. Based on these factors, he believes that

- There is a question unanswered.

- The research group can answer it.

- Getting the answer will be worth the effort.

There is a question unanswered means that a question remains from previous research, either the investigator's or someone else's. Every problem at hand has a scientific past—a context of prevailing beliefs and ideas that form the starting point and justification for further action. This context includes not only work accepted as correct, but also previous research viewed as incomplete or wrong. Here, knowing the scientific literature is crucial. Little is gained by solving a problem that already has been solved by others.

The research group can answer it means that adequate methodological, infrastructure, personnel, and financial resources are available to get the job done. Laboratories, field stations, and other research settings resemble small businesses. Resource limitations often prevent new initiatives from starting even if they would be worthwhile to carry out. Investing in one project almost always means that something else will not be accomplished.

Finally, *getting the answer will be worth the effort* means the investment of limited resources is justified. At one end of the scale are questions of broad scientific significance and interconnections within a research field—the more connections, the more meaning. At the other end of the scale are questions that are answerable but not very interesting—"So what!" questions.

What happens if the researcher's intuitions and beliefs turn out to be wrong? Possible consequences go beyond just not getting the answer or getting an answer viewed by others as unimportant. In the competitive world of science, failure can result in not getting one's research grant renewed or not getting tenure or not getting many other things that, taken together, effectively end one's career as a scientist. And it is not only about career. Failure also means lost opportunities to experience the emotional high of solving challenging scientific puzzles, being the first to learn something new, influencing how others think, and adding to the cumulative knowledge of science.

Once an investigator begins to pursue the answer to a problem at hand, inertia develops. When Srere invited me to join his laboratory, both he and I expected that I would use cultured rat liver cells to study cellular regulation of enzymes. Research about cell adhesion was not on the radar screen. I was to become part of his laboratory infrastructure. Simply noticing something unexpected was not sufficient reason to change direction. Research trainees often notice unexpected things—mostly irreproducible. Advisers try to keep trainees on course and prevent them from going off on unproductive tangents. Otherwise, progress will be delayed or halted altogether.

My situation was atypical. I was the first person in Srere's laboratory to use cell culture techniques. Consequently, I was in a sense establishing this aspect of the laboratory infrastructure. Because I was inexperienced, I had to read the literature to learn about the basics of cell culture. I became aware of the work of Saul Roseman, another biochemist and a friend of Srere's. Roseman recently had put forth a new hypothesis about the role of enzymes in cell recognition and cell-to-cell adhesion (20). After I showed Srere the results of my experiment and told him about Roseman's work, Srere called Roseman to find out if the enzymology of cell adhesion was a matter worth pursuing.

Cell adhesion is a fundamental feature of biological organisms. Adhesion holds us together—literally. Many human diseases, including cancer, exhibit pathologies that result from changes in cell adhesion. With Roseman's encouragement, Srere agreed that I could begin a new line of research in the laboratory. Separately, Roseman wrote to me, "The chemistry of this process is obviously unknown and any contribution that you make will undoubtedly be a major one" (21).

Pursuit

Regardless of whether the problem at hand has been under investigation for years or only recently initiated, pursuit requires that the investigator or group of investigators imagine or guess what might be the solution. With this guess in mind, the next experiment can be designed. The goal will be to confirm or reject one's guess. Unlike high school and college science laboratory, no textbook tells you what to do next. No instructor knows the answer in advance. Instead, pursuit turns out to be a little like playing hide and seek without knowing for sure what one is seeking or where the object has been hidden. To play means to enter into a dialog with the world. Claude Bernard, one of the mid-nineteenth-century founders of modern biomedical research, described this dialog as becoming *an inventor of phenomena* (22).

Answering even what appears to be a simple experimental question can present challenging obstacles. Imagine trying to determine the best way to drive between home and work during rush hour. Finding the answer requires critical prior decisions and commitments. What does "best" mean? How should "best" be measured? What routes should be compared? If best means fastest, then transit time from home to work would provide a good measure. If, instead of fastest, best means "most scenic," then what method should be used?

Other issues emerge: How should traffic accidents be counted? Even in the absence of traffic accidents, some day-to-day variation in transit time inevitably will emerge for a given route. How many drives along each route should be tested before reaching any conclusions?

Prior decisions and commitments not only will influence how the research is carried out, but also will determine the potential results that might be obtained. How many ways can you go fishing? Time of day, location, and type of fishing gear used will determine the range of fish likely to be caught. How many ways can you travel from Boston, Massachusetts, to Burlington, Vermont? Possibilities include airplane, bus, automobile, bicycle, and foot. The method chosen will establish limits regarding what can be seen along the way, where one can stop, and how long the trip is likely to take. Different methods might be better suited for a weekend outing to enjoy the New England fall foliage compared to a quick-turnaround business trip.

Expectations about how data should appear also limit possibilities for discovery. In 1913, archeologist Hans Reck discovered the first human skeleton at Olduvai in Africa. He also reported, surprisingly, that there were no stone tools in the area. Years later, Reck returned to Olduvai with fellow archeologist Louis Leakey. This was Leakey's first visit to Olduvai. Before leaving, Leakey bet Reck that they would find stone tools within 24 hours of arrival. Leakey won. Olduvai Gorge contained thousands of tools. Different rocks were available for making tools in Europe and East Africa. Reck had been looking for European-style Stone Age tools. On his previous trip to Olduvai, he had seen the rocks but did not notice them as tools (23).

Because experimental design and choice of methods limit potential experimental outcomes, more than one explanation can account for why experimental outcomes do not meet expectations and predictions. First possibility—wrong expectations or predictions.

Second possibility—wrong experimental design or methods. Further complicating matters, much of biomedical research depends on analysis of populations. Predictions about populations cannot be transferred easily to individuals. Overwhelming evidence links cigarette smoking with lung cancer. Many people develop lung cancer without smoking. Others have smoked and not developed the disease. Individual exceptions cannot disprove population-based conclusions.

The outcome of the Women's Heath Initiative (WHI) study on hormone replacement therapy that began in 1993 and involved thousands of women provides a good example of the difficulty in reaching definitive conclusions confirming or disproving one's expectations and predictions. Extensive research on different animal models carried out over a long period of time all led to the conclusion that hormone replacement therapy should protect postmenopausal woman against heart disease. Everyone was surprised, therefore, when the evidence began to show that WHI participants who received hormone replacement therapy were more likely, not less likely, to suffer a heart attack. As a result, WHI halted the research in July 2002. Was the prediction that hormone replacement therapy protects postmenopausal women wrong? Alternatively, was there something wrong with the design of the research? "Why didn't the effects seen in human beings match lab results? Had they been testing the wrong hormones or using inappropriate animal models all this time? Were the deficits in their own work, in the clinical studies, or in efforts to link the two?" (24).

Those who continued to believe that hormone replacement therapy might be beneficial for women raised critical issues about the study population and about the hormones used in the study.

We must be careful not to generalize broadly from these specific regimens [of the WHI] to the potential effects of other steroid analogs and treatment protocols. The effects of these steroids depend upon preparation, dose, sequence of administration, duration of postmenopausal hormone deficit, and tissue-specific and situation-dependent context of treatment. The observation that genetic background and health status of women at the time of treatment influence their responses adds yet another level of complexity that must be considered when designing studies, clinical trials, or prescribing hormone therapies. (25)

One aspect of continued follow-up work focused on women who were 50–59 years old at the time of enrollment in the WHI trial. These research subjects exhibited substantially decreased coronary artery calcification (26), consistent with the original expectations underlying the research. Therefore, depending on the subpopulation of women, the benefits of hormone replacement therapy may outweigh the risks.

The philosopher Karl Popper suggested that research advances by falsification of hypotheses (27). As the foregoing examples suggest, falsification can be difficult to achieve in everyday practice. Indeed, when scientists have good hypotheses, they do not give them up just because the data do not fit, at least not at first. Nevertheless, they are open to the possibility of being wrong. In everyday practice, Popper's idea of falsification signifies being open to the possibility of being wrong.

Control experiments

The discussion above focuses on the limitations of overall experimental design in determining possible outcomes. A second factor of equal importance concerns whether experimental methods are working as expected. "Don't waste clean thinking on dirty enzymes!" warned biochemist Efraim Racker (28). When I was a graduate student studying enzymology, I heard this expression regularly from my Ph.D. thesis adviser Jon Nishimura. We were trying to learn the mechanism of action of an enzyme called succinyl coenzyme A synthetase. If the enzyme was not sufficiently purified away from other contaminating enzymes, then we might mistake the properties of the contaminants for characteristics of the enzyme.

Control experiments function to test the adequacy of research design and methodology. In the experiment described in figure 2.3, no serum (condition 1) was a control for increasing amounts of serum (conditions 2–5). Standard medium (condition 6) was a control for medium with growth promoters (conditions 7–10) and growth inhibitors (conditions 11 and 12). All the conditions were tested using triplicate samples. I expected that the triplicates would show similar results. That the methodology appeared to be working as expected made the observation that arsenite inhibited cell stickiness stand out all the more. If there had been a lot of variability, then the arsenite finding would have seemed like just another unexplained variation.

Let me illustrate the importance of control experiments using a common example from everyday experience. Suppose I go on a diet. Two weeks later, I want to determine if my diet is working. I stand on my bathroom scale. I weigh the same. What are the possibilities? One, my diet is not working. Two, my scale is not working. If the scale were not working properly, then concluding anything about my diet would be meaningless.

So, what could be done to determine if the scale was working properly? Check other things whose weight I know? Check to make sure that the scale reads zero with nothing on it? Perhaps the scale needs adjustment. Only after I convince myself that the scale works as expected can I conclude reasonably that my weight has remained the same. Well, then perhaps my expectations are faulty. Maybe I have to wait at least three weeks before the effects of the diet will become noticeable. Of course, if what really concerns me is how I appear to others—not how much I weigh—then the scale cannot determine the success of my diet no matter what the scale reads. Rather, what really would count for data would be hearing from my friends, "Fred, you've lost weight! You look good."

Distinguishing data from noise

Even if the controls all work, one still can be left with uncertainty about how to interpret experimental data. In high school and college science experiments, the correct answers usually are known in advance. But in everyday practice of science, an investigator experiences an ambiguous field of experimental results. Only those results both noticed and seen as meaningful become data. The experimental drug UK-92,480 could not have been developed to treat penile dysfunction unless someone looking at the results not only took seriously the reported side effects, but also anticipated the drug's potential usefulness to reach a clinical end point different from the prevention of angina.

Data unnoticed or seen as irrelevant become equivalent to experimental noise. Not surprisingly, as one learns more about the experimental system under study, what counts for data or noise can change. A further description of my postdoctoral research with Paul Srere illustrates this point. Figure 2.4 shows the first experiment that I carried out in Srere's laboratory when I began my studies on cell adhesion. I measured the number of cells that stuck to a culture dish surface after different times up to 120 minutes. Studying the rate at which a reaction occurs was a typical thing for someone to do who was trained in enzymology. I had begun to treat the cell adhesion process as if adhesion were analogous to the formation of an enzyme–substrate complex.

I also photographed the cells through a microscope at the end of 2 hours and 24 hours. Figure 2.5 shows these photographs.

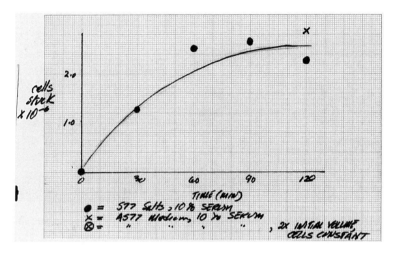

Figure 2.4. Graph Showing Cell Sticking as a Function of
Time—Data Equals Cell Number
From the Author's Notebook TC-I, p. 56, October 1970.

My training did not include photomicroscopy—the remarkably low quality of the images attests to this deficit. Yet, despite the presence of large and small pieces of dust and lack of proper focus, the cells were visible enough to count. In other words, what I understood as data was the number of cells that stuck to the dishes. Cell appearance was irrelevant. I continued this way of thinking even after leaving Srere's laboratory. In the 44 figures and 28 tables that I published in nine papers during 1971–1974, the data were all quantitative—number of cells sticking. Not a single photomicrograph or analysis of cellular morphology can be found in my papers.

An important aspect of the cell adhesion problem eluded all my efforts, namely, the identity of the biological glue for adhesion. Cell culture medium contains serum, a complex mixture of blood proteins. The biological glue likely was a protein in serum. However, the number of cells sticking to culture dishes was similar with or without

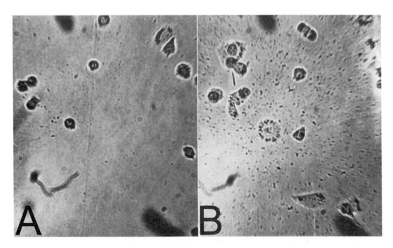

Figure 2.5. Photomicrographs Showing Cells After 2 hrs (A) and 24 hrs (B)
From the Author's Notebook TC-I, p. 59, October 1970.

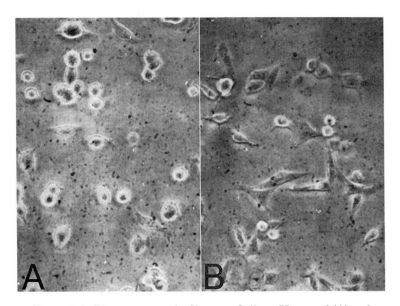

Figure 2.6. Photomicrographs Showing Cells on Untreated (A) and
Serum-Treated (B) Culture Dish Surfaces—Data Equals Cell Shape

From Grinnell, 1975. Cell Attachment to a Substratum and Cell Surface
Proteases. *Arch Biochem Biophys.* 169, 474–482, Figure 4.

serum in the medium. Figure 2.6 shows photomicrographs from an experiment in which serum was absent from the culture medium but was used to pretreat some of the culture dish surfaces. I noticed that the cells changed shape in the culture dishes that had been pretreated with serum (compare panel B with panel A) (29). I realized that cell shape change might be a better measure for the presence of the biological glue than cell sticking.

What happened between 1970 and 1975 that allowed me to think about cell shape as potential data? Perhaps the change occurred because I had started collaborating with an electron microscopist who was interested in cell structure. Perhaps it was because I began teaching histology to first-year medical students. In histology, microscopic features of cell size, shape, staining intensity, and location matter most. In any case, my intuition about what might count as data had changed.

Like figure 2.5, the images in figure 2.6 are of poor quality. They were good enough to show biochemists but not good enough for cell biologists. I had to learn photomicroscopic methods and to optimize experimental conditions to make the changes in cell shape more convincing. Figure 2.7 shows results from approximately the same experiment as in figure 2.6, but carried out six months later (30). Now there was no ambiguity in the results. Cell shape change (or cell spreading) instead of sticking became a useful assay for biological adhesion molecules. In the 57 figures and tables that can be found in my research papers published during 1975–1978, most show photomicrographs or measure changes in cell shape. The number of cells sticking no longer was of central importance. Indeed, if there were too many cells in the visual field, then I was unable to see cell shape clearly. What started as noise became data; what used to count for data became irrelevant.

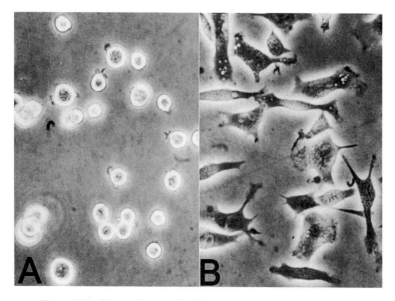

Figure 2.7. Photomicrographs Showing Cells on Untreated (A)
and Serum-Treated (B) Culture Dish Surfaces—
Data Equals Cell Shape (Refined)

From Grinnell, 1976. Cell Spreading Factor. Occurrence and Specificity of
Action. *Exp Cell Res.* 102, 51–62, Figure 2.

One final point—going back to figure 2.5, one can observe (using some imagination) that the cells changed shape between 2 hrs (A) and 24 hrs (B). I had the opportunity to begin to investigate shape change five years earlier. I missed the opportunity because I did not notice shape change as a feature worth studying.

Thought styles

Throughout this chapter, I emphasize the idea that an investigator's intuitions and expectations play a major role in everyday practice of science. These intuitions and expectations are part of a researcher's *thought style*, an idea developed in Ludwik Fleck's book *Genesis and Development of a Scientific Fact* (31). Fleck traces transformation of ideas about syphilis from fifteenth-century causes explained in terms of astrology and religion to the twentieth-century concept of infectious disease. He introduces the idea that every scientist has a unique thought style. The thought style depends on the person's education, experience, and temperament. Some aspects of thought styles are shared by many individuals and make up the prevailing beliefs of the research *thought community*. Other aspects uniquely reflect the person's biography and personality. A lot of similarity exists between Fleck's description of thought styles and the idea of paradigms later developed independently by Thomas Kuhn (32).

Interpersonal communication depends on overlapping thought styles. Groups of individuals who share common interests and activities (social, political, religious, scientific, etc.) tend to use a common language in carrying out group activities. Effective speakers communicate according to expectations of what sort of conversation would be appropriate to a particular situation. I discuss bioethics issues with diverse audiences ranging from high school students to postdoctoral fellows and residents—from community groups to hospital staff and biomedical research departments. In each case, I try to adjust the level of communication to the audience. If I use words or ideas outside the typical range of experience of audience members, then they likely will misunderstand me or misunderstand what I am trying to say. I will be wasting their time and mine.

In science, thought styles influence what researchers experience, what they think an experience means, and the subsequent actions they take. The thought style includes assumptions and beliefs about particular research problems. It includes the potential for noticing—Pasteur's famous "Chance favors the prepared mind!" It includes expectations underlying experimental design. It includes intuitions about what distinguishes data from noise. Even the anticipation that a particular problem might be solved through biomedical research depends on a person's thought style. Choosing to do research aimed at understanding and curing some disease means that the disease is viewed as a biomedical problem rather than a legal, religious, or educational problem (33).

Two pairs of competing experimental approaches play a central role in most researchers' thought styles. The first pair balances descriptive against mechanistic studies. Descriptive studies ask the question, "What's there?" Mechanistic studies ask questions such as, "How did it get there?" or "How does it work?" In biology, perhaps the most famous descriptive study was naturalist Charles Darwin's *Voyage of the Beagle.* Evolution by natural selection was the mechanistic answer that Darwin proposed to account for his observations about the diversity and distribution of organisms.

Until 150 years ago, descriptive studies characterized most of biology and biomedical research. Over the past 150 years, mechanistic work has become increasingly prominent. Understanding mechanisms enhances possibilities for intervention and modification of natural processes. Nobel Laureate Arthur Kornberg's description of biomedical research in the twentieth century highlights the mechanistic approach (34). When Nobel Laureate Robert Koch began his work on tuberculosis just before the turn of the twentieth century, bacteria were well known, but the relationship between bacteria and disease was poorly understood. Koch isolated bacteria from diseased organisms and showed that inoculating the bacteria into healthy organisms could cause the disease to occur. Following the guidelines established by Koch's overall experimental approach ("Koch's postulates"), investigators seeking to find the microbes responsible for a variety of other human diseases dominated biomedical research for the next two decades.

After the microbe hunters came the vitamin hunters, seeking to discover the causes of diseases such as scurvy and rickets. Dietary deficiencies required a mechanistic understanding reciprocal to infectious diseases. That is, absence, not presence, of something was responsible for causing the disease. Symptoms should appear after a suspected vitamin is removed from the diet and disappear after it is added back.

After the vitamin hunters came the enzyme hunters, whose goal was to identify the key enzymes responsible for cellular metabolism such as those of the citric acid cycle. Finally, the gene hunters arrived, trying to identify gene mutations that influence human susceptibility to diseases such as diabetes and cancer and that are closely linked to others such as cystic fibrosis and sickle cell disease.

Without knowing the genetic and protein composition of cells, understanding the genetic basis for disease and the complex networks that regulate cell function presents an insurmountable task. Consequently, major new descriptive studies were under way by the end of the twentieth century—the human genome and human proteome projects—with the goal to catalog all of the genes and proteins of humans and other organisms. Over time, descriptive and mechanistic studies feed off each other. Descriptive studies lead to new mechanistic questions. Mechanistic studies eventually require new levels of descriptive knowledge to advance to the next level.

The second pair of competing experimental approaches balances holistic against reductionist studies. Every biological system can be viewed either as an organized whole or in terms of its individual parts. Holistic studies focus on the organized whole. Reductionist studies focus on the individual parts.

Succinyl coenzyme A synthetase—the enzyme on which I worked as a graduate student—is composed of 20 amino acids. These 20 amino acids are organized into a specific sequence—about 350 amino acids long. Four subunits combine to form a single, intact enzyme molecule. Knowing the amino acid composition of an enzyme (reductionist approach) provides relatively little insight into its mechanism of action. Gaining insights into the mechanism of the enzyme requires knowing not only the amino acid composition, but also the amino acid sequence of the subunits and the organization of the subunits in three-dimensional space (holistic approach).

In the history of molecular genetics, researchers initially were skeptical that deoxyribonucleic acid (DNA) had sufficient complexity to be the source of genetic information. DNA has a relatively simple chemical composition—a linear sequence composed of four different nucleotides: thymidine, adenine, guanine, and cytosine. The key to understanding meaning depended on recognizing that embedded in the organization of DNA there was a *genetic code*. Read in three-letter nucleotide groups, the sequence has the potential to spell out 64 unique "words." These unique words provide instructions for mRNA translation into proteins: where to start, where to stop, and the order of amino acids to be inserted. With only a few modifications, the genetic code is uniform throughout all life forms.

Thought styles and persistence

Understanding the idea of thought styles helps explain the unique features of each investigator's approach to doing research even when those features run contrary to prevailing beliefs in the research community. Nobel Laureate Barbara McClintock expressed this idea explicitly in her Nobel banquet speech:

> I have been asked, notably by young investigators, just how I felt during the long period when my work was ignored, dismissed, or aroused frustration....My understanding of the phenomenon responsible for rapid changes in gene action...was much too radical for the time. A person would

need to have my experiences, or ones similar to them, to penetrate this barrier. (35)

Understanding the idea of thought styles also helps explain the ability of an investigator to persist in a particular line of research even when the research fails to go as expected. In chapter 1, I mentioned François Jacob's memoir and his comment about the research that led to the discovery of messenger RNA: "But nothing worked. We had tremendous technical problems. Full of energy and excitement, sure of the correctness of our hypothesis, we started our experiment over and over again" (36).

And here is what Nobel Laureate Rita Levi-Montalcini wrote regarding her experience in the fall of 1952 at the Institute of Biophysics in Rio de Janeiro, where she pursued the work that ultimately led to discovery of nerve growth factor, a key regulator of cell growth and development: "Even though I possessed no proof in favor of the hypothesis, in my secret heart of hearts, I was certain that the [cancerous] tumors that had been transplanted into the embryos would in fact stimulate [nerve] fiber growth" (37). The tumors did stimulate fiber growth but, unexpectedly, so did normal tissue fragments. This observation was

> the most severe blow to my enthusiasm that I could ever have suffered....After suffering the brunt of the initial shock at these results, in a partially unconscious way I began to apply what Alexander Luria, the Russian neuropsychologist has called "the law of disregard of negative information"...facts that fit into a preconceived hypothesis attract attention, are singled out and remembered. Facts that are contrary to it are disregarded, treated as exception, and forgotten. (37)

Levi-Montalcini ignored the result with normal cells and focused on purification of the factor from tumor cells. She was convinced that the factor was tumor specific. Later, when she knew more about the factor, she returned to study normal cells. Then she realized that nerve growth factor was of even broader biological significance than originally appreciated and functioned in the regulation of normal as well as malignant tissues.

Thought styles, gestalts, and schemes

Occasionally, I discuss practice of science with middle school or high school students visiting the University of Texas Southwestern Medical Center. I project figure 2.8 to help the students understand the idea of thought styles. "What do you see?" I ask. They write down their responses, which we then discuss. Everyone sees "C" as a smiley face. There is less agreement about "A" and "B," which appear to subsets of students as soccer balls and bowling balls, respectively. We discuss the idea that each student has a thought style, and that the thought style determines what the image appears to mean. Parts of an image take on different meaning according to how the whole image is interpreted. The dot in the center becomes part of the design of a soccer ball in "A," the thumbhole of a bowling bowl in "B," and the nose of the smiley face in "C."

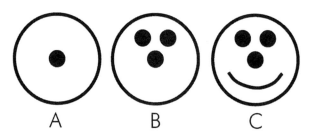

Figure 2.8. Gestalt Images

Psychologists call the reciprocal relationship in figure 2.8 between parts and wholes the *gestalt* quality of images. How the parts of a gestalt image appear depends on one's interpretation of the whole. Simultaneously, one's interpretation of the whole gives meaning to the parts (38). After thought style changes, things will no longer look the same. For a simple demonstration, we can return to figure 2.2. In trying to discern the out-of-place and comical element, each reader searches according to personal thought style. That thought style determines how to interpret the phrase "out-of-place and comical element," how to search through a complicated visual field, and whether searching is worth the effort. When first visualizing figure 2.2, the possibilities for observation are essentially limitless. Figure 2.9 shows the comical element magnified—a stick figure of a fisherman with a catch. Now the observer will be unable to resist visualizing the fisherman when looking at figure 2.2 in the future.

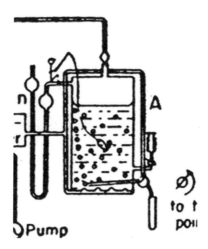

Figure 2.9. The Successful
Fisherman in Figure 2.2

Thought styles should be understood as dynamic structures. They are dynamic because they constantly change in response to new experiences. At first, they provide the context for new observations. Then they expand and differentiate through incorporation of the new observations. Then the expanded and differentiated thought style provides a new context that permits further observations, and so forth. One never returns to precisely the same place. Viewed over time, the circle of discovery (figure 1.1) appears as a spiral rather than a circle. Philosopher Patrick Heelan called this the hermeneutic spiral of science (39).

On a day-to-day basis, changes in one's thought style are very small. Tomorrow's experiments will be designed keeping in mind today's results. The work will appear continuous even though ideas and findings that appeared crucial at the beginning can later become irrelevant or even absurd. When a major change in the interpretation of the results does occur—moments of eureka or ah-ha—the meaning of past data itself changes. Over time, investigators learn and relearn the identity of what they are seeing. All discovery, wrote Fleck, is "supplement, development, and transformation of the thought style" (31).

Evolution of thought styles resembles the transformation of *schemata*. Schemata are the motor and conceptual structures proposed by psychologist Jean Piaget to explain how the child discovers the world. In Piaget's account, schemata first permit children to interact with (*assimilate*) their surroundings. Then the schemata become modified (*accommodate*) in response to those interactions. Experience is both made possible by and limited by schemata. A child moves a toy truck to himself by pulling a blanket on which the truck rests. This schema can be repeated, or the schema can be generalized to other movable objects (a ball). In addition, the schema can be generalized to other pulling devices (a string). Through experience, the child's schemata evolve, bifurcate, and develop into a matrix of practical and conceptual knowledge. The meaning of the world unfolds, a process that Piaget called *genetic epistemology* (40).

Encouraging evolution of thought style through collaboration and technological development

Childlike features of innocence and freshness play an important role in all creativity (41). Picasso commented that every child is an artist; the problem is how to remain an artist after one grows up. The challenge for every researcher is how to maintain a sense of innocence and freshness in the face of an evolving thought style that over time becomes increasingly dense and inflexible. Too much experience can be as much a hindrance as too little. Most researchers do their most creative work toward the early part of their careers (42).

"Men who have excessive faith in their theories or ideas," wrote Claude Bernard,

> are not only ill prepared for making discoveries; they also make very poor observations. Of necessity, they observe with a preconceived idea, and when they devise an experiment, they can see, in its results, only a confirmation of their theory. In this way they distort observations and often neglect very important facts because they do not further their aim. (22)

The thought style can act as a constraint and impediment to discovery when it influences investigators to see and think what they expect and constrains them from seeing and thinking more (Plato's paradox revisited).

When the artist Michelangelo looked at a block of marble, he knew based on his intuition and experience what figure was imprisoned within. He also knew precisely what sculpting tools and methods should be used to remove the excess stone and to release the figure. The results might have turned out differently if Michelangelo decided to try out a new tool with which he had no previous experience, or if he had given the job to one of his students.

When investigators look at blocks of reality, thought styles inform and guide their perception of what new features might be contained within and how they should be dis(un)covered. If a researcher sets to work with new tools, or delegates the research to trainees, or collaborates with other established scientists, then unique opportunities for discovery sometimes arise. New tools, new trainees, and new collaborators all increase opportunities to challenge thought styles, thereby promoting their transformation.

Historian Derek de Solla Price emphasized the relationship between new tools and discovery (43). Sometimes an investigator uses a new method or instrument to answer a specific question. And sometimes an investigator uses a new method or instrument because of its availability, hoping that something interesting might happen. Researchers call this latter approach a *fishing expedition*. They are hoping to see aspects of the universe never before observed—out of the realm of their thought styles. *I don't know what it is. I've never seen one before.* Such an outcome creates new questions and possibilities for future research.

Research aimed at making discoveries (*basic*) sometimes is distinguished from research aimed at transforming discoveries into new technologies (*applied*). In everyday practice of science, the dividing line is difficult to identify. Today's discoveries depend upon technologies developed last year (or sooner) just as much as next year's new technologies depend on today's discoveries. Conceptual understanding and methodological advances complement each other in ongoing cyclical fashion, presenting the researcher with novel domains for future research. Science really is an endless frontier.

Beyond new methods and tools, interacting with trainees and collaborators expands the thought community doing the research. A thought community exists

> whenever two or more people are actually exchanging thoughts. He is a poor observer who does not notice that a stimulating conversation between two persons soon creates a condition in which each utters thoughts he could not have been able to produce either by himself or in different company. (31)

Common features of their thought styles permit investigators to interact, while the differences lead them to observe and think about things in unique ways.

> Thoughts pass from one individual to another, each time a little transformed, for each individual can attach to them somewhat different associations. Strictly speaking, the receiver never understands the thought exactly in the way the transmitter intended it to be understood. After a series of such encounters, practically nothing is left of the original content. Whose thought is it that continues to circulate? (31)

In collaborative interactions, trainees, and other established scientists each bring unique qualities. Trainees will be more open to change because they have fewer prior commitments—more flexible thought styles. However, their influence potentially will be limited because they are expected to fit into the infrastructure established by the more senior person. Established scientists will be less open to change but more likely to notice subtle differences—experimental and conceptual—overlooked by those with less experience. When disagreements arise, trying to reach a consensus sometimes requires making explicit previously hidden assumptions of each other's thought styles.

Taken together, interpersonal relationships and interactions establish a thought community of mutual interests and critical feedback. Sociologist Howard Gardner identified these communal elements—mutual interests and critical feedback—as instrumental features of all creative accomplishment (41). Within thought communities, credibility happens to discovery claims. The credibility process is the subject of chapter 3.

3

CREDIBILITY

Validating Discovery Claims

Current work in my laboratory aims to identify the physiological signals and mechanisms that regulate wound healing in the skin. Changes in mechanical tension are thought to be one of the key signals associated with the wound healing process. A few years ago, one of the students in my laboratory discovered that in response to a rapid decrease in mechanical tension, skin cells called fibroblasts activate an important cell signaling pathway. We were very excited by her findings. We thought we had discovered a fundamental mechanism regulating events at the end of wound repair. In addition to publishing our findings, we presented the discovery at professional meetings and at other universities. Then I heard about research from another laboratory showing that the same cell signaling pathway could be activated if cells were wounded. Subsequently, we confirmed their interpretation. Ours had been wrong. Cell wounding—not a change in mechanical tension—was the principal cause responsible for modulation of the signaling pathway that we had observed.

Through the evolutionary process of discovery, scientists learn to see and think about the world differently than the world has ever been seen or thought of before. But what is the status of their new knowledge? We all act within our own unique life situations. Experience comes as *mine, here, now* (1). All discoveries include elements of subjectivity and ownership. No matter how diligently researchers try to be unbiased about their work, they cannot escape the potential for misinterpretation, error, and self-deception. Consequently, we should think of new discoveries initially as protoscience—discovery claims. To transform discovery claims into scientific discoveries, investigators turn to other scientists to establish the credibility of the work. Through the credibility process, an individual's *mine, here, now* becomes transformed into the community's *anyone, anywhere, anytime*—Professor Particular becomes Professor Anybody. The credibility process cannot be omitted. Those who make a discovery but ignore the credibility process are in the same position as those who never made the discovery at all.

Not all aspects of experience will be equally amenable to the credibility process. In chapter 2, I described the complexity of trying to figure out the best way to drive between home and work during rush hour. If we agree that best means fastest, then several different types of measuring devices might be used to determine the time required for the commute. If somebody says, "But it didn't feel that long," most of us will place higher trust in a stopwatch than in the feeling. If best means most scenic instead of fastest, then establishing credibility is much more difficult. Even if we take photographs, any scale or index of scenic appeal likely will be a highly personal matter. Perhaps you like panoramic views, whereas I prefer lakes and birds.

The scientist's struggle to learn new things about the world sometimes leads to spiritual intuitions about the meaning of the world or about the meaning of life (2). "It is humbling for me and awe-inspiring," commented Francis Collins, head of the U.S. Human Genome Project, at the White House Press Conference announcing completion of the rough draft of the genome in 2000, "to realize that we have caught the first glimpse of our own instruction book, previously known only to God" (3). "The most beautiful experience we can have," wrote Einstein, "is the mysterious. It is the fundamental emotion that stands at the cradle of true art and true science. Whoever does not know it and can no longer wonder, no longer marvel, is as good as dead, and his eyes are dimmed" (4).

In describing how he relates to the natural world, artist Joan Miró captured the sense of the mysterious:

> As I work on a canvass I fall in love with it, love that is born of slow understanding. Slow understanding of the nuances—concentrated—which the sun gives. Joy at learning to understand a tiny blade of grass in a landscape. Why belittle it?—a blade of glass is as enchanting as a tree or a mountain. (5)

Such personal existential moments cannot be tested for their scientific credibility, despite the profound impact they have on the individual.

Repeatability, continuity, and intersubjectivity

The possibility that there can be credible knowledge represents a fundamental belief of science. Expectations about credible knowledge are implicit in the sensibility of everyday life experience. Consider the following absurd conversation between Jones (J), who works in a coffee bar, and Smith (S), who has been showing up every morning at 9 A.M. for the past month and ordering the same thing (adapted from 6):

J: Hello! Good to see you again. The usual?

S: Usual what?

J: Nonfat, café mocha espresso with no whipped cream.

S: You're not making sense.

J: Well, then, what can I do for you today?

S: That's a long story—one day; I haven't time for it.

J: Something wrong?

S: Don't jump to conclusions; you might not make it, too much distance to cover.

J: If you'll excuse me, sir, I have some things to do. Perhaps I can serve you another time.

S: What other time? It will still be now. Now is when I want service.

Philosopher Maurice Natanson suggested that this conversation is absurd because it violates our expectations of repeatability and continuity. Repeatability means that what occurred in the past should be repeatable in the future. Continuity means that what takes place in the future should be a natural extension of what occurred in the past. The conversation also violates our expectation of intersubjectivity. Intersubjectivity refers to my experience that others with whom I share the world are similar to me in fundamental ways. If I were to exchange places with another person, I expect that she would see/hear/feel/smell/taste things similarly (albeit not identically) as I do.

Underlying all social interactions are the features of repeatability, continuity, and intersubjectivity (7). Science makes these expectations the basis for credible knowledge. Credible discoveries should be repeatable by the individual, continuous with previous scientific knowledge, and able to be verified and validated by others—anyone, anywhere, anytime.

The foregoing expectations resemble the basis for modern market research. When individuals in business or politics come up with new ideas about products or packages for anything from peanut butter to presidents, marketing experts test the new ideas on potential consumers in small groups (focus groups) or in small geographic regions (test markets). The responses of focus groups and test markets will be used to predict the likelihood of success with larger populations and to modify the original ideas to make them more appealing.

Objects of the marketplace can derive their meaning and significance entirely from social attitudes and cultural context. Think of hula hoops in the 1950s and pet rocks in the 1970s. Credibility in science aims for meaning and significance beyond the subjectivity of cultural context. This aim has its origins in our evolutionary past. Throughout history, the human attempt to overcome famine, disease, and other natural threats has depended on our ability to anticipate and modify future events based on those that already have occurred. In *Taking Darwin Seriously*, philosopher Michael Ruse imagines the following scene:

> One hominid arrives at the water hole, finding tiger-like footprints at the edge, bloodstains on the ground, growls and snarls and shrieks in the nearby undergrowth, and no other animals in sight. She reasons: "Tigers! Beware!" And she flees. The second hominid arrives at the water, notices all of the signs, but concludes that since all of the evidence is circumstantial nothing can be proven. "Tigers are just a theory, not a fact." He settles down for a good long drink. Which of these two hominids was your ancestor? (8)

In evolutionary history, the test of credibility is nothing less than survival. And science aims for nothing less.

It is one thing to aim for meaning and significance that transcend subjective cultural context, but quite another to achieve this goal. Successful application of science to real-world problems through technology has established the overall credibility of scientific thinking. We live in a world heavily dependent on science and technology. Along the way, however, much research has been driven by the social and political biases of investigators and the cultures in which they live. The combination of diverse and competing interests of researchers and their freedom to dissent or concur openly offers the best protection against research becoming mere marketing.

The credibility process: An overview

Figure 3.1 presents an overview of the credibility process through which discovery claims put forth by individual researchers and research groups become transformed into the research community's credible discoveries. If significant enough, these discoveries become

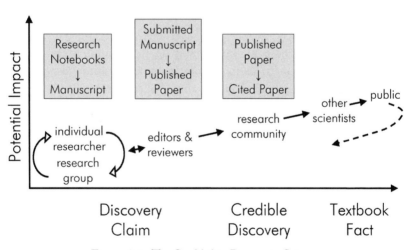

Figure 3.1. The Credibility Process in Science

the textbook facts of science education. As a discovery claim works its way through the credibility process, the knowledge of the discovery reaches a larger and larger potential audience. Having a larger potential audience increases the potential impact of the work. Researchers who think their work is sufficiently important can try to increase its impact by alternative mechanisms such as press releases. Having a press release can influence the visibility of a discovery claim but in no way substitutes for the credibility process.

Figure 3.1 shows that the credibility process involves a complex social structure of researchers interacting with each other. Within this social structure, one finds insiders and outsiders with respect to each discovery claim. Insiders correspond to the circle of researcher, research group, and collaborators making the claim. Outsiders correspond to editors and reviewers, the larger research community, other scientists, and the public. At every level, one finds shared and competing thought styles about diverse aspects of research, including previously accepted observations and facts, appropriate methodological approaches, and important problems requiring further investigation. Thought styles are useful to describe communities as well as individuals. Interaction and confrontation of individual and community thought styles become the dialectic of the credibility process. Discovery claims made credible by this dialectic will be incorporated into and refashion the prevailing thought style. The more important the discovery, the greater will be the refashioning. For the community as for the individual, *all discovery is supplement, development, and transformation of the thought style* (9).

The dialectic of creativity in the arts follows a similar pattern:

One begins with a set of individuals of varying abilities, talents, and proclivities, each engaged in work in a particular domain. At any historical moment, that domain features its own rules, structures, and practices [thought style], within which they are expected to operate.... The works (and the workers) so judged [as creative] come to occupy the most important spot in the dialectic: they actually cause a refashioning of the domain. The next generation of students, or talents, now works in a domain that is different, courtesy of the achievements of highly creative individuals. And in this manner the dialectic of creativity continues. (10)

Only after scientific discovery claims become public can other researchers in the field learn the details of the new findings and use them (or not) as suits their own purposes. Therefore, making a discovery claim public represents a transformational point of the credibility process (11). Modern scientific publication began with the *Philosophical Transactions of the Royal Society.* The Royal Society was founded in London in 1660—the first of the world's modern scientific academies. Scientists wanted to establish the priority of their discoveries before sharing the results with others. An independent journal could fulfill this need. Publication of an article in *Philosophical Transactions* (and subsequent journals) became the "on-the-record, validated public statement of the claims made by its authors, like a witness statement under oath in the court of scientific opinion" (12). Reports of discoveries by all other formats were viewed as informal and tentative.

Discovery claims that are used and turn out to be useful will increase in credibility over time. The process is highly dynamic—not a static event—and corresponds to philosopher William James's pragmatic conception of truth:

> Credible discoveries are those that we can assimilate, validate, corroborate, and verify.... The credibility of a discovery is not a stagnant property inherent in it. Credibility happens to a discovery. It becomes credible, is made credible by events.... Its verity is the process of veri-*fication*. Its validity is the process of valid-*ation*. (paraphrased from 13; italics original)

Transformation of research notebooks into manuscripts

As indicated in figure 3.1, transformation of research notebooks into manuscripts occurs at the level of the circle of insiders. The size of the insider group can vary and has changed during recent years. Almost all papers published in 1974 in the prestigious biomedical research journal *Cell* had either one or two authors, and only two papers had more than four authors. Individual researchers tended to work alone or in small collaborative groups. Since then, research groups have become larger and collaborations more common. By 2007, one- or two-author papers in *Cell* were rare, and papers with five to ten authors were common. Notwithstanding this increase, most discovery-oriented biomedical research still can be characterized as small-scale science. Small-scale science has limited objectives and usually is carried out under the direction of a single principal (lead) investigator or co-investigators who are located at one or two research institutions.

Some research objectives are so broad and multidisciplinary—for example, the Superconducting Super Collider Project and the Human Genome Project—that they cannot be managed by small-scale science and require a more elaborate organizational framework. Large-scale science can involve many principal investigators, institutions, and nations (14). Papers published in 2001 in *Science* and *Nature* that reported the initial sequencing of the human genome had hundreds of authors (15, 16). At the time when funding for the Superconducting Super Collider Project was discontinued in 1993, approximately 2,000 scientists and engineers were working on the project at more than 200 institutions worldwide (17). Notwithstanding the scope of large-scale science, discovery claims that emerge from the research will still be subject to the credibility process.

Researchers within the circle of insiders decide if experiments have succeeded or failed, what counts as data, and what the data mean. Every experiment stands alone in terms of design, execution, and interpretation. Every experiment also represents a conceptual and technical continuation of and response to what has gone before. Conclusions are continuously under revision, influenced by new experiments, by discussions between researchers, and by discovery claims published by other laboratories.

Moving a discovery claim beyond the circle of insiders in figure 3.1 requires transformation of raw notebook data into tables, figures, and other images for presentation. By notebook, I mean research records regardless whether they are written, photographic, electronic, and so forth. Only a small percentage of raw notebook data will be so transformed. Often, many preliminary efforts are required before investigators understand their project sufficiently well to carry out successful experiments. Early results tend to be incomplete and sometimes irreproducible. Much of the work represents false starts, dead ends, and methodological failures. Even heuristic experiments from which one learns new and important information usually need to be repeated and modified. After the experimental system and research question become clear, then experiments can be designed that are consistent, complete, and stylistically appropriate to show to outsiders. These become the demonstrative experiments found in research papers (9).

In most academic research, transformation of research notebooks into manuscripts will be carried out under the control of individual investigators along with their group members and collaborators. Usually, investigators want to make public their new research findings as quickly as feasible. If findings have possible commercial value, such as the discovery of potential therapeutic drugs or devices, then publication sometimes will be delayed to allow time for patent applications to be filed to protect the monetary value of a discovery.

Other research oversight arrangements also are possible. In laboratories controlled by the pharmaceutical industry, decisions about what to publish frequently are determined by managers as well as individual investigators. Data arising from the research sometimes will be designated as *proprietary*, that is, private and not to be made pubic, even if the data might be of value to the research community.

Talking about science

Transformation of the contents of research notebooks into tables, figures, and other images typically requires a lot of talking among insiders and between insiders and outsiders. This ongoing chatter occurs in labs, offices, halls, e-mail, blogs, and telephone conversations. Frequently, members of one or several laboratories get together for works-in-progress discussions. Disagreement or confusion about the meaning of the research findings leads to more experiments, more talk, more experiments, more talk, and so on.

Insiders also get opportunities to present (test market) discovery claims to outsiders at departmental seminars and national and international conferences. Such preliminary presentations not only elicit supportive or skeptical comments but also increase the spread of information about new discoveries before the claims are published formally.

Comments by others can have a profound impact on the direction of one's research. In his memoir, Sir Peter Medawar wrote that his early ideas were subjected to "scathing" criticism by his professor— "That would be quite a nice little experiment, wouldn't it, and that's just what you really like, isn't it—little experiments?" (18). Medawar admitted to himself the truth of this comment and was determined to overcome it. Subsequently, he began his work on burn wound repair, skin grafting, and finally the immunology of transplantation, research for which he was awarded the Nobel Prize.

Similarly, physicist Freeman Dyson wrote that one of the big turning points in his life resulted from a meeting with Nobel Laureate Enrico Fermi in the spring of 1953:

> In a few minutes, Fermi politely but ruthlessly demolished a programme of research that my students and I had been pursuing for several years. He probably saved us from several more years of fruitless wandering along a road that was leading nowhere. I am eternally grateful to him for destroying our illusions and telling us the bitter truth. (19)

The above examples show how talking about science can change an investigator's overall research approach. Even more frequently, as illustrated by the following two examples from my own experience, key comments and questions can play a critical role in determining how work on a particular research project evolves.

In 1977, I presented results of our research on cell adhesion in a seminar at the University of Tennessee. In the discussion period after my talk, a researcher in the audience named Andrew Kang suggested that one of our methods was flawed. We believed our data showed that an important biological adhesion protein called *fibronectin* was required for cells to attach to *collagen*. Collagen is a major structural protein of many tissues. Kang remarked that by preparing collagen as a dried film, we had destroyed its normal structure. Instead, he said that we should have used the collagen in its native three-dimensional form. Later, he told me what experimental methods could be used to prepare native collagen. When we repeated our experiments following his advice, we got exactly the opposite results from the original findings. That is, cell attachment to native collagen did *not* require fibronectin. Kang's critique of our work was especially important because most investigators in the field at that time ignored the impact of native versus denatured collagen structure on cell behavior.

The second example occurred about a decade later. My laboratory had begun to carry out clinical research studies to learn whether fibronectin might be useful as a topical therapy for patients with non-healing leg ulcers. Nonhealing leg ulcers have become a major health problem, especially for the elderly. When I presented a talk about fibronectin and wound healing at the Tarpon Springs Tissue Repair Symposium in 1987, I mentioned our preliminary findings regarding fibronectin as a therapeutic treatment. Someone in the audience asked about the condition of fibronectin in the ulcers.

When I returned home from the symposium, I mentioned the question about fibronectin in chronic ulcers to Annette Wysocki, a postdoctoral fellow working with me on the project. She suggested that we cover the patients' ulcers with a watertight dressing and analyze the wound fluid that accumulated beneath the dressing. Watertight dressings were relatively new then, although now they can be found in drugstores and supermarkets. Wysocki knew about them because of her nursing background. We were able to collect samples from several patients and discovered that much of the fibronectin in the chronic wound environment was in a degraded form. Subsequent work from our laboratory and others showed that high levels of protein-degrading enzymes are a general feature of chronic skin ulcers.

Formalizing discovery claims: Research papers

As research and talk about research proceed together, discovery claims achieve repeatability, continuity, and intersubjectivity within the circle of insiders. The group's thought style evolves toward a consensus view regarding methodology, results, and significance of the findings. Sociologists and anthropologists studying everyday practice of science call this interaction the social construction or manufacture of knowledge (20, 21).

Once formalized, discovery claims are made public in highly stylized format. Research papers begin with an introduction section that places the discovery claim in its historical scientific context. The goal of the introduction is to connect the new findings with those reported previously by the laboratory and by other researchers. After the introduction, the results section presents the tables, figures, and other images derived from the notebooks. Finally, a discussion section explains and interprets the results and summarizes the key conclusions that extend previous knowledge. A methods section inserted before the results or after the discussion explains how the experiments were accomplished, and a reference section at the end of the paper contains citations to other research papers relevant to the discovery claim.

Typically, the introduction and results sections will present a logical and internally consistent account of the studies. This account can be quite different from the historical events that actually occurred. By the time a series of experiments are complete, the rationale behind the work may have changed, earlier thinking discarded, and older findings reinterpreted in light of later ones. These contrasting accounts are well illustrated by comparing sections from François Jacob's description of the discovery of mRNA taken from his memoir *The Statue Within* with the *Nature* paper in which the discovery claim was formalized. First the memoir:

> Our confidence crumbled...we found ourselves lying limply on a beach, vacantly gazing at the huge waves of the Pacific crashing onto the sand. Only a few days were left before the inevitable end. But should we keep on? What was the use? Better to cut our losses and return home....From time to time, one of us repeated the litany of failed manipulations, trying to spot the flaw....Suddenly, Sydney gives a shout. He leaps up, yelling, "The magnesium! It's the magnesium!"

Immediately we get in Hildegaard's car and race to the lab to run the experiment one last time.... Sydney had been right. It was indeed the magnesium that gave the ribosomes their cohesion. But the usual quantities were insufficient.... This time we added plenty of magnesium. (22)

And now the paper:

The bulk of the RNA synthesized after infection is found in the ribosome fraction, provided that the extraction is carried out in 0.01 M magnesium ions[12].... Lowering of the magnesium concentration in the gradient, or dialyzing the particles against low magnesium, produces a decrease of the B band and an increase of the A band. At the same time, the radioactive RNA leaves the B band to appear at the bottom of the gradient. (23)

In the paper, the beach and Sydney's shout are gone, along with the rest of the adventure. The way the results are written makes it appear that the experiments were carried out from the beginning with high (0.01 M) magnesium, as recommended previously by others (superscript citing reference 12). Then, the findings with low magnesium are described as if they were done as intentional controls—not a summer of frustration—to show that removal of high magnesium caused the complexes to fall apart, that is, "the radioactive RNA leaves the B band" (23).

Data selection and historical reconstruction are inevitable for most research papers. As logic replaces history, the plot of every scientific paper becomes the scientific method. In his memoir, Jacob says:

Writing a paper is to substitute order for the disorder and agitation that animate life in the laboratory.... To replace the real order of events and discoveries by what appears as the logical order, the one that should have been followed if the conclusions were known from the start. (22)

It cannot be otherwise. Research papers that contained all of the data and historical events relating to a discovery claim would overwhelm everyday practice of science with useless information. The credibility process is concerned with verification and validation of demonstrative data and conclusions, not with preliminary results and adventures. Sir Karl Popper made this point explicit in *The Logic of Scientific Discovery*:

> The initial stage, the act of conceiving or inventing a theory seems to me neither to call for logical analysis nor to be susceptible of it. The question how it happens that a new idea occurs to a man—whether it is a musical theme, a dramatic conflict, or a scientific theory—may be of great interest to empirical psychology; but it is irrelevant to the logical analysis of scientific knowledge. This latter is concerned...only with questions of justification or validity. (24)

One commentator suggested that Popper's book could have been called *The Logic of the Finished Research Report* (25).

Although the section of research papers that describes methods and materials provides important details regarding how the experiments were carried out, the complexity of experimental technique often requires personal exchanges between investigators to clarify just how things actually were done. My favorite example is a story from the history of transmission electron microscopy (TEM). Beginning around 1940, TEM was introduced to observe biological samples at very high magnification (26). The samples were placed on thin plastic films. Researchers prepared these films with a plastic called Formvar. The method for preparing Formvar films involved a series of steps: Immerse a glass microscope slide in Formvar solution. Remove the slide and allow the Formvar to air dry. Cut the edges of the dried Formvar film with a razor. Slowly immerse the Formvar-coated glass slide in a container of water, allowing the Formvar film to release from the glass slide and float on the surface of the water.

For many investigators, the last step did not work—the Formvar films would not release from the glass slides. According to microscopy legend, a leading research center invited an investigator who had published research using Formvar-coated grids to visit and present a lecture about his work. While he was there, the investigator was asked to demonstrate his method for preparing the films. Just before immersing the glass slide with dried Formvar film into the container of water, he held the slide a few inches in front of his mouth and breathed out—"huff!" The small amount of water vapor deposited on the Formvar film by this "huff" made releasing the film from the slide easier. Subsequently, the huff step became widely known and a part of standard practice.

Gatekeepers and peer review

Before researchers can formally present their discovery claims to the scientific community, they have to get by the gatekeepers. The gatekeepers—scientific journal editors and reviewers—evaluate submitted manuscripts and determine if a paper should be published, revised, or rejected. Since journal editors and reviewers also are members of the scientific community, the gatekeeper process in science exemplifies what is called *peer review*. The attitude of the gatekeepers toward discovery claims can be summed up by the inscription on the coat of arms of the Royal Society, *Nullius in verba*, which Sir Peter Medawar translated as, "Don't take anybody's word for it."

The peer review process provides published papers with an initial stamp of authenticity. Papers published without peer review lose this important element of legitimization. How seriously the community takes the legitimizing function of peer review can be appreciated by the controversy over a 2004 paper on the subject of intelligent design that appeared in *Proceedings of the Biological Society of Washington* (27). (I discuss intelligent design in some detail in chapter 6.) Shortly after publication, the Council of the Biological Society of Washington released the following statement:

> The paper was published at the discretion of the former editor, Richard V. Sternberg. Contrary to typical editorial practices, the paper was published without review by any associate editor; Sternberg handled the entire review process. The Council, which includes officers, elected councilors, and past presidents, and the associate editors would have deemed the paper inappropriate for the pages of the *Proceedings*.... [T]he Meyer paper does not meet the scientific standards of the *Proceedings*. (28)

Reading the council statement, one is reminded of a judge instructing jury members to disregard evidence with which they have been presented. Science and the judiciary share interesting similarities related to credibility because each requires mechanisms for determining who gets to present information (publish or testify), how one gets to present the information, and how the presentation should be evaluated.

Comparison with the judicial system helps clarify peer review in science. When trials involve complex technical issues, expert witnesses often will be hired by one side or the other or appointed by the court. These experts provide testimony—sometimes contradictory—regarding theories or techniques relevant to the facts of a case. Judges are responsible for ensuring that the testimony of the experts will increase the potential for jury members to understand the facts correctly. If a judge errs by preventing expert witness testimony that should have been allowed or by allowing testimony that should have been prevented, then the possibility for a fair trial may be compromised.

In court cases that specifically involve scientific matters, the U.S. Supreme Court decision in the 1993 case of *Daubert v. Merrell Dow Pharmaceuticals, Inc.* established that

> the trial judge must make a preliminary assessment of whether the testimony's underlying reasoning or methodology is scientifically valid and properly can be applied to the facts at issue. Many considerations will bear on the inquiry, including whether the theory or technique in question can be (and has been) tested, whether it has been subjected to peer review and publication, its known or potential error rate, and the existence and maintenance of standards controlling its operation, and whether it has attracted widespread acceptance within a relevant scientific community. The inquiry is a flexible one, and its focus must be solely on principles and methodology, not on the conclusions that they generate. (29)

The court guidelines resemble those used by journal editors and reviewers of scientific papers, perhaps restated as questions:

1. Are the techniques appropriate?

2. Could any scientist potentially have done the work?

3. Are the results interpreted in an appropriate fashion?

4. Are the studies reasonable in light of ideas previously accepted by the community?

Experts in the judicial framework frequently provide the same testimony case after case. Repetition is the rule rather than the exception—each trial has a separate identity. Unlike the judicial situation, scientific journal editors and reviewers are looking for something beyond repetition, namely, novelty. What most impresses gatekeepers in scientific review are new ways of thinking. If the work presented in a research paper simply confirms somebody else's discovery claims, then it will be of less interest. Scientific journals develop reputations depending on their ability to attract and publish the most novel work with the greatest impact on the research community. The more successful a journal, the more selective it can be regarding which papers to publish.

Of course, scientific journal editors and reviewers sometimes make mistakes in evaluating the importance of discovery claims. Some well-known examples include rejection by *Science* of the radioimmunoassay method (30), rejection by *Cell* of site-directed mutagenesis (31), and rejection by *Nature* of monoclonal antibodies—although the journal published a shortened version of the original paper as a letter (32). Each of these discoveries was important enough that its discoverer received a Nobel Prize for the work.

The social structure of credibility

Judges and expert witnesses rarely are part of the same thought communities. In science, editors, reviewers, and researchers not only share the same thought community, but also constantly interchange positions. Figure 3.2 shows the overall dynamic. The principal investigator (PI) publishes research papers. These papers help the PI obtain research grants (solid arrows) and help attract trainees to work in the PI's laboratory (dotted arrows).

Researchers who succeed as principal investigators gain in personal credibility—*credit-ability*. Prior success feeds back to help subsequent submissions (dotted arrows). In his evolutionary account of science, philosopher David Hull described credit-ability in terms of the likelihood that a scientist's contributions would be used by and

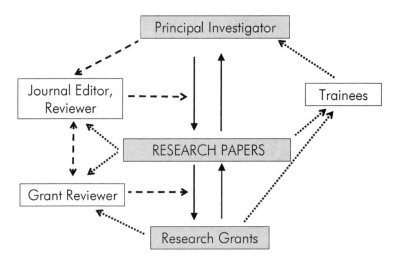

Figure 3.2. The Social Structure of Credibility
Different Types of Arrows Are Used to Emphasize Different Interactions
as Described in the Text.

have an impact on subsequent research and researchers (33). This
impact occurs directly through successful discovery claims and indirectly through the training of future principal investigators. Impact
also occurs when principal investigators act as community representatives to evaluate the credibility of work (papers and grants) submitted by others (dashed arrows).

Some people argue that there is too much potential for intellectual inbreeding in the community structure depicted in figure 3.2.
So much interconnectedness makes conflicts of interest inevitable.
Manuscripts and grant proposals that I am asked to review sometimes come from friends or former students. If there is a potential
for conflict of interest, then I disclose the relationship to whoever
is responsible for oversight of the review process. The response that
frequently comes back: "Thanks for your candor, Fred. We'd accept

your comments as long as you don't feel that your connection with __ interferes with your judgment. If your scientific friendship with __ makes you feel uncomfortable in providing a recommendation, then don't do it." In the end, the decision is mine to make.

The social structure of credibility requires a lot of trust. Being knowledgeable about a subject places peer reviewers in the best position to offer an informed opinion about a paper or grant. At the same time, peer reviewers also are in the best position to take advantage of the advanced knowledge received through the review process. Therefore, functioning of the credibility process requires that researchers not use each other's submitted work inappropriately to advance their own ends. Despite the challenges of conflict of interest and potential for misuse of advance knowledge, the rigor of peer review frequently receives credit for much of the success of contemporary science.

Research grants

Figure 3.2 includes research grants as part of the credibility process. Doing research requires obtaining grants or some other source of funding. In addition to submitting manuscripts, the circle of insiders shown in figure 3.1 also submits research grant proposals. Grant proposals are subjected to a credibility process different from that of discovery claims because the research has yet to be accomplished. Reviewers of grant proposals submitted to the National Institutes of Health (NIH) are asked to consider five key points (34):

1. Significance: Does this proposal address an important problem?

2. Approach: Are the conceptual or clinical framework, design, methods, and analyses adequately developed, well integrated, well reasoned, and appropriate to the aims of the project?

3. Innovation: Is the project original and innovative?

4. Investigator: Is the investigator appropriately trained and well suited to carry out this work?

5. Environment: Does the scientific environment in which the work will be done contribute to the probability of success?

These criteria show that for a grant to be judged favorably, reviewers need to be convinced that there is a question unanswered, that those within the circle of insiders have the capacity to answer the question, and that getting the answer will be worth the effort. In short, every aspect of an investigator's thought style becomes the focus of review.

Even if a grant proposal appears worthwhile, requests for funding far exceed agency resources. Although applications originate with and follow the interests and ideas of those within the innermost circle, success often depends on the ability to link interests and ideas with research funding priorities. Research funding priorities are established by public interest groups and the federal government. As indicated by the dashed line in figure 3.1, credibility is not a unidirectional process.

The credibility process goes on and on

At the end of a trial, the jury votes. Afterward, jury members rarely challenge the testimony or the outcome of the trial even if they do not like the results of the vote. The challenge option is open to the lawyers, but their focus usually will be on matters of procedure that led to the outcome. Of course, new evidence sometimes becomes available later and leads to new trials, such as the introduction of DNA testing as a forensic method.

In contrast to the judicial outcome, there is never an end to trial by science. A vote by the scientific community does not settle the truth of a matter after a discovery claim becomes public. Researchers, unlike jury members, carry out additional experimental work to assimilate, validate, corroborate, and verify matters of substance and methodology. Credibility in science inevitably remains open. What appears credible one day may be viewed as flawed the next, and vice versa. The trial goes on and on.

Going "on and on" is the basis for the self-correcting feature of science. Even if editors and reviewers accept flawed work, then its flawed character can be detected by other researchers. "It must be possible for an empirical scientific system to be refuted by experience," wrote Popper (24). Potential for self-correction—what Popper called *falsifiability*—is one of the most powerful reasons that scientific knowledge becomes highly reliable. The "cash value" of falsifiability is that science gives up the possibility of unchangeable truth—Truth with a capital "T"—and settles for credibility.

When other investigators within the research community test the credibility of a discovery claim, they typically do so either by trying to incorporate the new findings in their own ongoing studies or by using the findings as the basis for developing a new line of research. One way of determining the research community's response to a discovery claim is by tracking the frequency with which papers describing discovery claims are cited in the subsequent research literature. The ISI Web of Knowledge tracks citations, and these data frequently are used to evaluate individuals, departments, universities, and even different countries around the world as a measure of success in scientific research (35).

The average citation rate for a paper published in the most prestigious journals in the field of cell biology is approximately 15/year. In addition to the number of citations, the pattern of citations reveals the dynamics of credibility. Figure 3.3 shows year-by-year citation

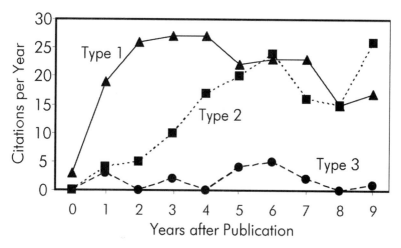

Figure 3.3. The Credibility Process and Impact

data for three of my papers that were published in the *Journal of Investigative Dermatology* (36–38). The three patterns in figure 3.3 illustrate three different responses of the research community:

- Type 1. Immediate impact. The research community finds the discovery claim useful soon after the claim is reported.

- Type 2. Delayed impact. The research community finds the discovery claim useful but only after a delay.

- Type 3. Incremental impact. The research community finds the discovery claim only of minor value.

Type 1 papers begin to refashion the prevailing thought style of the research community as soon as they are published. Their credibility is established quickly. Type 2 papers also help refashion the prevailing thought style of the research community but not until the community catches up. Type 3 papers have little impact on the community thought style. Although researchers would like all of their work to have the impact of type 1 papers, much of what is published falls into the type 3 category. One cannot predict in advance how a research project will turn out. However, the dynamic interrelationships shown in figure 3.2 make publishing papers essential even if what one has learned is of only incremental importance.

Challenging the prevailing thought style

Nobel Laureate Albert Szent-Györgyi's prescription for discovery was *seeing what everybody else has seen and thinking what nobody else has thought.* René Magritte's 1936 oil painting *Perspicacity* shows a seated artist staring at a solitary egg on a draped table. On his canvas, he paints a bird in full flight. As expressed in Szent-Györgyi's prescription and Magritte's painting, discovery frequently requires unconventional thinking. The more novel a discovery claim, the greater its potential to refashion the thought style and affect subsequent research in the field. At the same time, novelty also challenges intersubjectivity and can come into conflict with the prevailing thought style. As a result, highly novel discovery claims sometimes are received with considerable skepticism by the research community.

I learned firsthand the difficulty of challenging the prevailing thought style when I was a postdoctoral fellow with Paul Srere. Srere was engaged in a dispute with many in the research community concerning how best to understand enzyme regulation. His focus was on the enzymes of the citric acid cycle. Many of these enzymes, he emphasized, were present in cells at much higher concentrations compared to those typically studied by biochemists. "But a cell is not a test tube!" Srere argued that understanding enzyme regulation required analysis of enzymes in the context of their supramolecular organization within cells (39).

Frustrated by the resistance of conventional thinking in the research community to his new ideas, Srere would begin his seminars by describing the *I Ching*. The *I Ching* is a Chinese philosophical system attributed to the Emperor Fu Hsi (circa 2800 B.C.E.) that is used to understand the past and predict the future. Paul explained to his audience (mockingly) that he was skeptical about the 5,000-year-old method until he examined the trigram symbols arranged in a circle. He was amazed to learn that Fu Hsi had predicted the citric acid cycle well before the cycle was discovered by Sir Hans Krebs. Then Srere would show the front cover of the Beatles 1969 album *Abbey Road*. He would point out some of the clues on the cover indicating that Paul McCartney had died, a popular conspiracy theory at the time. Srere told his listeners that he was fascinated "by the evidence, logic and inevitability of the conclusion that McCartney died.... It was a compelling story. The only flaw was that he was alive." And then Srere reached his timeless conclusion: "Given a large mass of data, we can by judicious selection construct perfectly plausible unassailable theories—all of which, some of which, or none of which may be right" (40).

While the skepticism of conventional thinking can delay acceptance of new ideas, if the ideas are correct, then the research community eventually catches up. One sign of the research community catching up with Paul Srere was the establishment in 1987 of an ongoing research conference titled Enzyme Organization and Cell Function (later renamed Macromolecular Organization and Cell Function). In recognition of his contributions to the field, Srere's fellow biochemists established the "Paul Srere Memorial Lecture" beginning with the 2004 conference.

The history of Nobel Prizes includes many examples of novel discoveries that were either ignored or disputed for years, for example, tumor viruses (Nobel Prize in 1966), chemiosmotic theory (Nobel Prize in 1978), transposable genetic elements (Nobel Prize in 1983), and catalytic ribonucleic acid (Nobel Prize in 1989). The presentation speech for Krebs when he won the 1953 Nobel Prize for the discovery of the citric acid cycle reminded the audience that "in the beginning Krebs was quite alone with his idea, and when he first presented it, it was criticized by many" (41). The presentation speech for Stanley Prusiner when he won the 1997 Nobel Prize for the discovery of prions contained similar remarks:

> The hypothesis that prions are able to replicate without a genome and to cause disease violated all conventional conceptions and during the 1980s was severely criticized. For more than 10 years, Stanley Prusiner fought an uneven battle against overwhelming opposition. (42)

In the end, accommodation of the thought style to novelty frequently follows the path described by William James: "First, you know, a new theory is attacked as absurd; then it is admitted to be true, but obvious and insignificant; finally, it is seen to be so important that its adversaries claim that they themselves discovered it" (13).

Error

Commenting on one of his competitors' papers, a nineteenth-century scientist wrote, "The paper contains much that is new and much that is true, but the new is not true and the true is not new" (quoted in 43). Error is a common and inevitable part of science. The "discovery" of polywater remains my favorite example of error. I was a graduate student when a letter published in *Nature* titled "'Anomalous' Water" sounded a doomsday alarm:

> Sir,— A report on the properties of "anomalous" water appeared recently in *Nature* (222, 159; 1969). The probable

structure of this phase was reported by Lippincott et al., who refer to the phase as polywater, a term descriptive of the structure.... After being convinced of the existence of polywater, I am not easily persuaded that it is not dangerous. The consequences of being wrong about this matter are so serious that only positive evidence that there is no danger would be acceptable. Only the existence of natural (ambient) mechanisms which depolymerize the material would prove its safety. Until such mechanisms are known to exist, I regard the polymer as the most dangerous material on earth. (44)

Polywater was scary stuff because of its potential to begin an unstoppable reaction whose outcome would destroy normal water and eventually all of life. Moreover, although the work reporting the structure of polywater came from a U.S. laboratory (45), the Russians made the original discovery and were doing most of the research on the topic. This was the time of the Cold War. Could the Russian scientists be trusted? I remember discussing the danger with other students. If they tried to destroy our water system, then eventually their water probably would be destroyed as well. What if they made a mistake? All life might end!

Not everyone was convinced of the danger. Another letter appeared in *Science* a year later titled "'Polywater' Is Hard to Swallow."

Proponents of polywater in the pages of *Science* and elsewhere may be interested to learn why some of us find their product hard to swallow. One reason is that we are skeptical about the contents of a container whose label bears a novel name but no clear description of the contents. Another is that we are suspicious of the nature of an allegedly pure liquid that can be prepared only by certain persons in such a strange way. (46)

Subsequently, new findings challenged the earlier work: "Thus we must conclude that all polywater is polycrap and that the American scientists have been wasting their time studying this subject unless, of course, it can be defined as a topic of water pollution and waste disposal" (quoted in 47).

Boris Derjaguin, who reviewed polywater for *Scientific American* in 1970, retracted the discovery in a 1973 letter to *Nature*: "Consequently, the anomalous properties of condensates may be explained, not by the formation of a new modification of water, as was previously supposed," but by unsuspected microcontamination of the samples (48). When asked how much polywater had been obtained in all, Derjaguin responded, "About enough for fifteen articles" (quoted in 47).

In the case of polywater, the source of the error was discovered. Sometimes, the source of erroneous research cannot be identified. On several occasions during my research career, observations that we made over a period of months suddenly and inexplicably could no longer be reproduced. We never discovered why. Most researchers have similar experiences during their careers.

Unfortunately, another reason that research reported by one laboratory sometimes cannot be validated, corroborated, and verified by others is that the research was bogus to begin with. Some scientists make up their experiments instead of actually doing them. Some scientists do the experiments but intentionally select or modify the results to prove a particular point. Such behavior is called *research misconduct*. Research misconduct and its close cousin, conflict of interest, threaten the trust that underlies practice of science. In chapter 4, I discuss these subjects as my focus moves from science to science and society.

II
SCIENCE AND
SOCIETY

4

INTEGRITY

From Science Policy to Responsible Conduct of Research

In part I, I discussed the central activities of everyday practice of science—discovery and credibility. Here I broaden the focus to encompass issues concerning science and society. Several years ago, I served on an Institute of Medicine panel whose charge was to analyze research integrity in relationship to the research environment. Our committee emphasized the need to understand everyday practice of science in the broader context of the surrounding sociocultural, political, and economic environment (1). Therefore, I begin this chapter by describing how the surrounding environment influences everyday practice.

The diagram in figure 4.1 illustrates (albeit in a simplified manner) the complex and hierarchical organization that characterizes the integration of science and society. This organization spans the range from researchers to institutions to the government to the public. Expectations based on accomplishments lead to questions about the future: What are we doing right? What are we doing wrong?

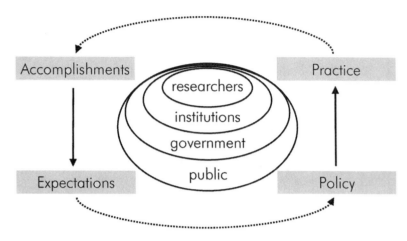

Figure 4.1. Science and Society

What more can we hope for? Should some research not be allowed? Expectations elicit government regulatory decisions and research funding allocations, collectively called *science policy*. Science policy plays an important role in determining every aspect of future practice, including what work will be done, who will do the work, and how the work will be financed.

Human embryonic stem cell research policy

The debate in the United States regarding human embryonic stem cell research illustrates the complex interactions between science policy and everyday practice of science. President George W. Bush's first prime-time televised speech to the American public took place in 2001 and addressed whether federal funds should be used to carry out research with cells derived from human embryos:

> Good evening. I appreciate your giving me a few minutes of your time tonight so I can discuss with you a complex and

difficult issue, an issue that is one of the most profound of our time. The issue of research involving stem cells derived from human embryos is increasingly the subject of a national debate and dinner table discussions. The issue is confronted every day in laboratories as scientists ponder the ethical ramifications of their work. It is agonized over by parents and many couples as they try to have children, or to save children already born.

The United States has a long and proud record of leading the world toward advances in science and medicine that improve human life. And the United States has a long and proud record of upholding the highest standards of ethics as we expand the limits of science and knowledge. (2)

President Bush went on to announce a policy that would not "sanction or encourage further destruction of human embryos that have at least the potential for life."

Most funding for biomedical research in the United States comes from congressional allocations to federal agencies. Therefore, President Bush's decision restrained the potential national effort to carry out human embryonic stem cell research. Societal groups with a stake in expansion of stem cell research responded by trying to influence political decision making at local, state, and national levels. Patient advocacy organizations argued that human embryonic stem cell research would lead to advances in therapy to treat the diseases about which they were concerned. Biomedical scientist groups argued that the research would lead to advances in understanding of disease onset and progression. The biotechnology industry argued that the research would advance the national economy. All three groups emphasized that the potential benefits of embryonic stem cell research would be delayed without federal funding. The United States might lose its opportunity to lead this important scientific field.

In the language of ethics, the arguments in favor of permitting destruction of human embryos to advance embryonic stem cell research are called *utilitarian*. Utilitarian arguments focus on the consequences of an action rather than who carries out the action or the nature of the action itself. Utilitarian arguments had sufficient political clout that five years after establishing his policy, President Bush was forced to cast his first veto to prevent passage of congressional legislation that would have overturned the federal funding ban.

Notwithstanding political relevance, utilitarian arguments do not address the issue of embryo personhood. President Bush and other opponents of destroying human embryos for research purposes (which includes the Catholic Church) hold an *essentialist* view of the embryo. They think of life as a continuum that begins as soon as the potential for a human being exists—an embryo is a person not yet born.

The Helsinki Code of Ethics represents the most widely established international set of ethical considerations to guide human research. Paragraph A5 of the Helsinki Code states: "In medical research on human subjects, considerations related to the well-being of the human subject should take precedence over the interests of science and society" (3). If embryos were persons, then carrying out embryo research would be unethical based on Helsinki A5 because such research would be sacrificing "persons" for the good of science and society. Utilitarian arguments would be irrelevant despite the potential benefits of stem cell research to patient care, biomedical science, or the national economy.

But are embryos persons? Many would argue otherwise. The human genetic identity of the clump of cells that makes up the embryo does not by itself establish personhood. Personhood can be understood as a quality of the embryo that emerges during development or at birth. Because a primitive body plan cannot be detected until day 14 of human embryonic development, day 14 often has been suggested as the earliest possible point of emergence.

If one accepts the viewpoint that embryos are not persons, then Helsinki A5 no longer is relevant. Utilitarian arguments now apply. Even allowing the utilitarian perspective, other ethical issues still require consideration and resolution. For instance, the U.S. National Academies recommended that researchers should not pay donors for embryos from which stem cells were to be derived. Such payments might result in exploitation of economically vulnerable couples and encourage increased commodification of human body parts (4).

In the absence of specific congressional legislation, President Bush had the power to restrict how federal funds were used. But he had no control over state funds. California voters, convinced overall of the appropriateness and benefits of human embryonic stem cell research, passed Proposition 71 by a 3:2 margin in the 2004 elections to establish the California Institute for Regenerative Medicine. Other states followed with their own initiatives. Private funding of the research continued, as well. In the absence of a national consensus or national regulatory policy, the ethics of destroying human embryos for embryonic stem cell research in the United States depends on whose money one uses to carry out the work.

The discovery by researchers of how to reprogram normal human cells to become pluripotent and similar in many ways to human embryonic stem cells without the involvement of embryos may resolve some of the political debate about human embryonic stem cell research (5). One imagines that supporters of President Bush's decision will credit the rapid research advances in cellular reprogramming to the wisdom of his policy. However, the question regarding appropriateness of research that results in destruction of human embryos remains unresolved. Indeed, many people do not realize that even if President Bush had permitted federal funds to be used to study stem cells derived from human embryos, the derivation of these cells using federal funds would itself still be illegal because of a congressional amendment barring use of federal funds for research that results in harm to human embryos (6).

Researchers in England discovered that human ova could be fertilized *in vitro* (IVF), and the first IVF human baby was born in 1978. Subsequently, most U.S. expert panels evaluating IVF recommended that embryo research should be carried out to make IVF a more safe and effective medical procedure. Their advice never became policy. Shortly after the discovery of IVF, the 1973 U.S. Supreme Court *Roe v. Wade* decision recognized abortion as a matter of the mother's personal privacy—her choice. Abortion and destruction of embryos became linked. Those who opposed *Roe v. Wade* also opposed any research that resulted in the death of human embryos. Subsequently, the anti-abortion lobby in the United States succeeded in blocking federal funding for embryo research. Ironically, one unintended consequence of the lack of research to improve the safety and effectiveness of IVF has been the need to prepare excess embryos for implantation. Those that have not been used—now more than a half million—are stored in the freezers of U.S. fertility clinics. More frozen embryos continue to accumulate. The availability of frozen embryos has become a major factor driving the debate regarding use of human embryos for embryonic stem cell research.

Science in the public interest

Whatever their views about human embryonic stem cell research, most people would accept President Bush's statement that advances in science and medicine can improve human life. The National Institutes of Health (NIH) states its mission as "science in pursuit of fundamental knowledge about the nature and behavior of living systems and the application of that knowledge to extend healthy life and reduce the burdens of illness and disability" (7).

The importance of coupling advances in science and technology with human well-being has been articulated in the United States since colonial times. America's first scientific hero, Benjamin Franklin, founded the American Philosophical Society in 1743. The inaugural volume of the society's *Transactions* offers this vision:

> When speculative truths are reduced [to technology], when theories grounded upon experiments, are applied to common purposes of life; and when, by these, agriculture is improved, trade enlarged, the arts of living made more easy and comfortable, and, of course, the happiness of mankind promoted; knowledge then becomes really useful. (quoted in 8)

Unfortunately, technological advances that appear to improve human life sometimes are accompanied by unintended and undesirable consequences with the opposite effect. The optimistic view implied by the 1939 DuPont company slogan "Better Things for Better Living...Through Chemistry" came under attack in the 1960s. Marine biologist Rachel Carson published *Silent Spring*, exposing the risk of pesticides and sparking the modern environmental movement. A decade later, economist Ezra Mishan labeled the automobile "the greatest disaster to have befallen mankind":

> Almost every principle of architectural harmony has been perverted in the vain struggle to keep the mounting volume of motorized traffic moving through our cities, towns, resorts, hamlets, and of course, through our rapidly expanding suburbs. Clamor, dust, fume, congestion, and visual distraction are the predominant features in all our built-up areas....Whether we are in Paris, Chicago, Tokyo, Düsseldorf, or Milan, it is the choking din and the endless movement of motorized traffic that dominate the scene. (9)

Those who are pessimistic about technology could make a long list of unintended, undesirable consequences, including depletion of the ozone layer in the earth's atmosphere and global warming. Most recently, concerns have developed about nanotechnology and nanomaterials.

> The convergence of nanotechnology with biotechnology and with information and cognitive technologies may provide such dramatically different technology products that the...development of policies and regulations to protect human health and the environment, may prove to be a daunting task. (10)

Notwithstanding unintended, undesirable consequences, the world has become an increasingly technological place. Governments believe that the continued well-being of their nations requires new scientific discoveries and technological development combined with high-level science education. In the United States, this opinion has been articulated in many national reports and proposals. In his 2006 State of the Union address, President George W. Bush introduced the American Competitiveness Initiative:

> America's economic strength and global leadership depend in large measure on our Nation's ability to generate and harness the latest in scientific and technological developments and to apply these developments to real world applications. These applications are fueled by: scientific research...a strong education system...and an environment that encourages entrepreneurship, risk taking, and innovative thinking. (11)

Government-funded research is believed to be a crucial component of modern technological innovation and development, but those responsible for science policy decision making want to know how much is necessary. They want to know how the impact of government investment can be maximized. The frontiers of science may be endless, but public resources are limited.

Attempts to link the scientific research agenda with specific national goals rarely lead to consensus. Consider the relationship between biomedical research and national health care. The United States spends more on biomedical research than any other country in the world. We also have the costliest health care system in the world. Yet our public health statistics are only average compared with other industrial countries. So, is there something wrong with the national investment in biomedical research? Not enough? Too much? Wrongly focused? Poorly accomplished?

Of the federal agencies that support biomedical research in the United States, the NIH provides the majority of the funding, currently about $30 billion/year. Congress wants to know what NIH accomplishes with the money it receives. Representatives of patient advocacy organizations want to know why more money is not spent on their diseases. Individual researchers want to know why there is not more money, period!

The NIH director regularly tries to answer such questions and criticisms through congressional testimony and in public interviews. In addition, NIH periodically produces research updates, such as *Investments, Progress, and Plans* (12) and *Doubling Accomplishments— Selected Examples* (13).

Funding First is a pro-science initiative sponsored by the Mary Woodard Lasker Charitable Trust. Its goal is to determine "the true economic value of our national investment in medical research" (14). The truly exceptional worth of investment in medical research is shown by comparing the costs of research with the economic benefits of reduced illness and increased life expectancy. More investment would be better.

A different type of answer comes from economists who argue that the success of biomedical research and development of new medical technologies is itself an important cause of rising health care costs:

> Rapid scientific advance always raises expenditures, even as it lowers prices. Those who think otherwise need only turn their historical eyes to automobiles, airplanes, television, and computers. In each case, massive technological advance drove down the price of services, but total outlays soared. (15)

At the final meeting of his advisory committee in 1999, former NIH director Harold Varmus voiced the concern that

> unless ways are found to reduce the cost of new therapies and diagnostics—or to pay the cost for all comers—the genomic-medicine revolution will bring more, deeper divisions between those who have access to high-technology care and those who don't. (16)

Dan Sarewitz suggests in *Frontiers of Illusion* that when politicians or scientists propose "more research" as the solution to social problems, they are using science as a "surrogate for social action" and avoiding the tough political, social, and economic decisions necessary to solve the problems directly. Maybe the reason that we have the costliest health care system in the world is the lack of political will to create a single-payer national health care system. Health care costs and public health statistics are social issues. More biomedical research is not going to solve the problem that almost 50 million Americans lack health insurance and ready access to health care. Sarewitz advises political and cultural institutions that if they really want to advance the public good, then they should act "as if scientific and technological progress had come to an end and the only recourse left to humanity was to depend upon itself" (17).

Rather than expect a direct linkage, we would do well to understand the relationship between practice of science and national goals as articulated in Vannevar Bush's founding vision of American science policy, *Science—The Endless Frontier*, written after World War II. Investment in scientific research is necessary, but research alone is not sufficient.

> Science, by itself, provides no panacea for individual, social, and economic ills. It can be effective in the national welfare only as a member of a team, whether the conditions be peace or war. But without scientific progress, no amount of achievement in other directions can insure our health, prosperity, and security as a nation in the modern world. (18)

Ambiguity of outcomes

Ambiguity inherent in everyday practice of science complicates the possibility of linkage between the research agenda and specific national goals. The outcome of specific research projects cannot be foreseen clearly, nor can we anticipate when discoveries made in one context will turn out to be highly useful for entirely different purposes. Again, *Science—The Endless Frontier* offers a clear understanding:

> One of the peculiarities of basic science is the variety of paths that lead to productive advance. Many of the most important discoveries have come as a result of experiments undertaken with very different purposes in mind. Statistically, it is certain that important and highly useful discoveries will result from some fraction of the undertakings of basic science; but the results of any one particular investigation cannot be predicted with accuracy. (18)

Rather than ambiguity, government officials seek account-ability. The Government Performance and Results Act (GPRA) passed by Congress in 1993 requires government agencies, including those involved in funding basic research, to establish short-term performance goals that are objective, quantifiable, and measurable. Responsibility for oversight of GPRA and related management tools rests with the U.S. Office of Management and Budget (OMB). On their Web site, OMB ranks NIH performance as *effective* (the highest rating possible) and states, "The program is achieving its goals." They also report, "The program's budget and performance integration is making progress in presenting the tie between funding requests and expected performance results" (19). How tightly can funding requests be tied to performance results? At best, one might identify fields that are most likely to contribute to solving particular problems. Beyond that, asking the unpredictable to be predicted and focusing on short-term achievements rather than long-term impact has the potential to distort and undermine basic research by discouraging innovation and risk taking.

A National Academies committee proposed that instead of establishing quantitative, short-term performance measures and expected performance results, programs funding basic research should be evaluated by expert panel review according to broad, qualitative criteria. These criteria would include comparison with other work being carried out in the field, consideration of whether the research is at the forefront of scientific and technological knowledge, appropriateness of research to agency goals, and potential value if successful. "Expert review is widely applied—used, for example, by congressional committees, by other professions, by industry boards, and throughout the realm of science and engineering—to answer complex questions through consultation with expert advisers" (20).

Subjectivity of expert review

By its very nature, expert review tends to be subjective and difficult to quantify. Moreover, because members of expert review panels are appointed, their work will be viewed with confidence only if panel members are selected to obtain balance in knowledge, interests, and perspectives. This is particularly important for panels appointed by government agencies. Criticism by diverse groups has been aimed at what has been called "the Bush administration's war on the laboratory" (21). Many commentators both in and out of politics claim that President George W. Bush's administration attempted more so than its predecessors to manipulate science policy by unbalanced appointment of individuals committed to the administration's social, political, and religious agendas.

Besides avoiding political manipulation, the problem of avoiding conflict of interest also presents a challenge to expert review panels. Financial interest is the most common source of conflict of interest, but institutional and ideological commitments are just as important. Previously made public statements or strongly held positions on a subject frequently reflect potential biases. Potential biases do not necessarily disqualify individuals from participating in expert committees, but their disclosure is essential to achieving balance.

Selecting balanced expert panels can be especially difficult if a consensus develops among scientists regarding extrascientific issues associated with a particular subject. For instance, is destruction of human embryos for embryonic stem cell research the morally correct thing to do? Most scientists now agree, but public support is only somewhat favorable (22). If scientists share a policy perspective in disproportionate homogeneity compared with the population at large, then appointing a committee that appears adequately balanced to those within and outside of science might be impossible.

The U.S. Food and Drug Administration (FDA) in particular has come under continual attack because of alleged financial conflicts of interest and lack of balance in the composition of its expert review panels (23). One highly publicized example involved Merck & Co., Inc., which received FDA approval in 1999 to market Vioxx, the blockbuster anti-inflammatory medication. Vioxx was recalled in 2004 because of adverse cardiovascular side effects. In between, sales of the drug reached billions of dollars. When an FDA advisory panel met in February 2005 to reconsider the safety of Vioxx and related drugs, the committee members voted in favor—17 to 15—to allow continued marketing of Vioxx notwithstanding its potential adverse effects. According to a *New York Times* analysis, the vote would have opposed continued marketing—14 to 8—if members with conflicts of interest were excluded (24).

Multiple desirable outcomes

Besides ambiguity of practice and subjectivity of expert review panels, differences of opinion regarding the desirable outcomes of the scientific research agenda present a major difficulty in linking everyday practice of science with national goals. A report titled *Federally Funded Research: Decisions for a Decade* by the Office of Technology Assessment discusses a range of different, desirable outcomes (25). At the top of the list of outcome measures is the research itself. One way that success of the research can be measured is by determining productivity and impact—the number of research papers published and patents filed and the frequency with which published papers and patents are cited in subsequent works (26).

Ironically, productivity in industry and in science can have different meanings and consequences. In industry, increased productivity means obtaining increased output for the same cost or less. In science, increased productivity does not necessarily take cost into consideration. Productivity can be increased by getting another research grant to add staff and equipment to the laboratory rather than by making existing staff and equipment more effective. Consequently, whereas competition in industry tends to drive costs down, competition in science sometimes has just the opposite effect.

A different outcome by which to judge the research agenda might be education of the current and future work force. In this case, indicators of success could be how many students are trained, how long the students take to get their advanced degrees, or what sorts of positions they move to after completing their education. As more time, effort, and funds are invested in educational activities, less investment will find its way to the research itself.

Yet another outcome might be increased institutional capacity. The greatest amount of funding for academic research and development tends to be concentrated in a relatively small number of colleges, universities, and research centers. Historically underfunded institutions often have trouble competing if scientific capacity is viewed as the primary basis for decision making regarding research funding. If incremental scientific capacity or increased research opportunities for students were the basis for decision making, then the underfunded institutions would be more competitive. Increased funding would have a bigger relative impact on the underfunded institutions compared with those already receiving the most research dollars.

The potential for local economic development also could be used to assess the outcome of the scientific research agenda. Just as a small number of academic institutions and research centers receive most federal funding for doing research, a few states and regions within states currently are responsible for most scientific research and development work. Here again, the underdeveloped areas will compete better for research dollars if potential impact is judged according to incremental rather than absolute criteria.

Paying attention to multiple outcomes is beginning to become formalized as part of applications to some government research funding agencies. Scientists who write research proposals to the National Science Foundation are required to discuss not only the intellectual merits of their work, but also the broader potential impact of their work on society:

> How well does the activity advance discovery and understanding while promoting teaching, training, and learning? How well does the proposed activity broaden the participation of underrepresented groups (e.g., gender, ethnicity, disability, geographic, etc.)? To what extent will it enhance the infrastructure for research and education, such as facilities, instrumentation, networks, and partnerships? Will the results be disseminated broadly to enhance scientific and technological understanding? What may be the benefits of the proposed activity to society? (27)

Finally, following the adage that "all politics is local" and recognizing the impact of science funding on local, regional, and state economic development, the most desirable outcome for many members of Congress is to bypass the expert review process and have federal funds directly allocated for specific science research projects within their political regions. This strategy, known as *earmarking*, always has been a part of politics but has increased dramatically for the sciences since 1980. According to an analysis carried out by the American Association for the Advancement of Science, earmarked funds requested in the 2007 federal budget accounted for 20–25% of the scientific research portfolios of some federal agencies (28).

Money policy

Up to this point in this chapter, I have emphasized matters of science policy that provide the broad societal context for everyday practice of science. From this point forward, I increasingly focus on individual researchers. For those of us doing science, money policy—how much money will be spent on research and how the money will be distributed—is the science policy decision with the most immediate and decisive impact on the research environment. Issues such as the integrity of researchers, trust, and conflict of interest can be understood best against the backdrop of money policy.

The biomedical science community has expanded dramatically in recent decades because of the increased federal investment in biomedical research. Over much of the same period, funding remained relatively constant or even declined for other key areas of science such as physics, chemistry, mathematics, and computer science. Less funding available means fewer research opportunities and fewer research positions.

"What's going to happen to the NIH budget" is the ongoing concern of the academic biomedical researcher. I started to hear about the "NIH budget crisis" in 1967 when, as a graduate student, I attended my first annual meeting of the Federation of American Societies for Experimental Biology (FASEB). In the years before the gambling casinos arrived, the FASEB societies used to meet together in Atlantic City, New Jersey. A group of us had gathered at the bar of the Strand Motel with our department chair, Alton Meister, at the center of attention. Someone asked Meister about funding. "Are things going to get better, Al?" "No," he said, "worse!"

Meister's "worse" has turned out to be mostly a matter of relative deprivation. To the academic world outside of biomedical research, figure 4.2 shows remarkable growth. Congressional budget allocations

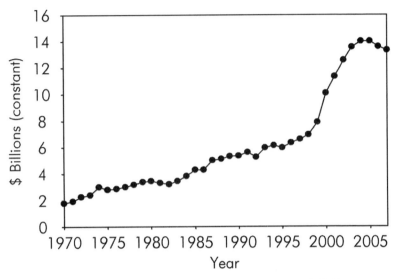

Figure 4.2. NIH-Funded Research at Universities and Colleges

for NIH permitted an almost eightfold increase (constant $) in support for academic research since 1970. Despite these overall increases, there were years during every decade when funds declined. The decrease in the early 1990s was widely expected to signal a long-term freeze in federal support for biomedical research. Yet in 1998, a five-year doubling of the NIH budget began. Following the doubling, the NIH budget entered an unprecedented period of decline (29).

Funding declines create tremendous turmoil in the biomedical community because they interfere with planning and raise concerns for anyone thinking about a career in the field. If funds remain stationary or decrease from one year to the next, but the number of competing investigators remains the same or increases, then the possibility of succeeding at grant winning becomes less favorable. In 2000, there were 16,827 new and 4,995 renewal applications submitted for funding. The success rate was 25.9% for new applications and 49.8% for renewals. Six years later, 22,148 new and 6,666 renewal applications were submitted, and the success rates were down to 16.3% and 33.5% (30).

"We are in the midst of an era of plummeting pay lines at the NIH," wrote the editor of the prestigious *Journal of Clinical Investigation* in a 2004 editorial titled "The Economy of Science":

> History shows that when the federal deficit is high, NIH pay lines tend to fall, and the impact on biomedical research can be disastrous....How many bright young people will be turned away from careers in biomedical research? How much innovative science will be delayed or never initiated, how many new cures never realized?...In the early 1990s, when obtaining funding from the NIH for biomedical research was generally regarded as a Herculean effort due to low pay lines...there was a generalized depression amongst biomedical researchers. (31)

The comment in the editorial about "generalized depression amongst biomedical researchers" reflects in part the *soft money* research environment in the United States. In a soft money research environment, salaries come largely from research grants. Before 1960, the federal government provided little support for faculty salaries. This changed after the President's Science Advisory Committee (PSAC) recommended that steps be taken to double the number of academic centers of excellence in the United States. The committee wanted to increase the number of first-rate academic centers—research universities—from 15–20 to 30–40 over the following 15 years. Toward this end, PSAC recommended federal support for university faculty:

> Since the federal government has a deep interest in the rapid increase in the quality and quantity of the nation's teaching

scientists, its agencies should in general, seek forms of support for basic research and graduate education, which will permit universities to enlarge their permanent faculties. (32)

Expansion succeeded beyond PSAC expectations. By 2007, NIH was providing at least some research-related funding to several thousand U.S. institutions. The institution at the top of the list received around $600 million (30).

When PSAC recommended the use of federal funds to enlarge university faculties, committee members envisioned that block grants would be made directly to universities and issued a caveat. "We recognize...the need for avoiding situations in which a professor becomes partly or wholly responsible for raising his own salary." But this caveat was lost along the way. Instead, what has evolved is a system in which success in winning research grant funding sometimes determines whether newcomers get faculty appointments, whether professors in the early stages of their careers keep their faculty appointments, and whether those of us with tenure keep our salaries even if our faculty appointments are guaranteed (33).

Academic researchers experience three kinds of conflicts in the soft money research environment, each creating unique problems. First, scientists are at risk of financial conflict of interest if one takes seriously the U.S. Public Health Service (PHS) definition of a significant financial interest as $10,000 or more. Second, researchers often experience conflicts of commitment trying to balance the demands of their universities with those of the expert panels that review and score the scientific merit of research grant applications. Satisfying the review panel can become more important to an investigator than satisfying the university where the research is carried out. Dedication to teaching, clinical practice, and other university service sometimes suffers as a consequence. Finally, a potential conflict occurs regarding mentorship. When graduate students and postdoctoral fellows are supported by research grants rather than by training programs, then they become laboratory employees as well as trainees. Laboratory directors have a fiduciary responsibility to their grants to ensure that employees carry out the proposed research. They also have a mentorship responsibility to ensure that trainees receive the education and experience necessary for future success. Sometimes, what is in the best interests of a trainee may not be in the best interests of the laboratory. For instance, good mentoring occasionally requires allowing a trainee to struggle with a problem and fail, which certainly does not increase laboratory productivity.

Trust

As funding success rates go down, researchers become more sensitized to potential problems with the grant review process. The NIH panel that reviews my grant applications currently has about 20 regular members. In a typical review cycle, the review panel will discuss 50 or so applications over a 2-day period. The discussions for each application will focus on the comments of several reviewers, each of whom analyzes multiple applications. While I require a couple of months to write a grant application, the reviewers likely will spend no more than several hours evaluating my proposal and preparing their comments. Finally, the review panel likely will spend no more than 15–30 minutes discussing and deciding the scientific merit of my proposal.

Expert review of grant applications is truly peer review. According to the NIH Web site, review panel members must be recognized authorities in their fields and must be funded investigators on projects comparable to those under review (34). NIH administrators organizing a review panel face a challenging task:

> On the one hand, peer reviewers should be informed on the subject of the proposal in order to judge it fairly. On the other

hand, the more closely a reviewer's expertise matches an applicant's, the more likely it is that the two could be direct competitors or allies. As one NIH scientific review administrator [put it], "We walk a fine line to get qualified reviewers without a conflict of interest." Although there are few documented cases of this, an opportunistic reviewer could sabotage a competitor's proposal or unfairly assist a friend. This tension is inherent in the peer review process. (35)

The idea that an opportunistic reviewer might sabotage a proposal frequently worries researchers when they submit their grant applications. Certain types of comments can influence the tendency to score an application slightly higher or lower in a subtle but significant way. Saying something like "Professor Particular's work is very thorough even though somewhat descriptive" will not help Professor Particular's interests. The "descriptive" label is a negative connotation often hard to shake. By contrast, describing someone's research approach as "mechanistic" would be positive. Small differences in scoring the scientific merit of a proposal can have a big impact on funding.

In the end, researchers have little choice except to trust that reviewers will treat them fairly, not only with respect to grant evaluation, but also with regard to the information that reviewers learn from the contents of the applications. The same experience and interests that make a researcher qualified to offer an informed opinion about a grant application also make the person likely to benefit from knowing the preliminary results and experimental thinking in the application. Likewise, scientists who act as peer reviewers of research manuscripts submitted for publication gain advance knowledge about new discovery claims. Although the contents of grant applications and submitted manuscripts are privileged and confidential information, reviewers cannot avoid being affected by what they have learned. New knowledge inevitably becomes part of one's future thinking. I like to draw an analogy between the peer review process and a bizarre version of playing poker—you show your competitors your cards—and then hope that they will continue to play their own cards more or less as if they had never seen yours.

The importance of trust goes beyond the review process to permeate every facet of practice of science. Trust is a fundamental principle and prior commitment necessary for functioning of the contemporary research environment. Trust begins in the laboratory. I can review the notebooks kept by my students and postdoctoral fellows, observe their actions, and discuss their findings with them. Eventually, I will have to trust that they are doing what they say they are doing and getting the results that they describe. When the research program involves more than one laboratory, there is even less possibility for oversight. The requirement for trust increases along with the physical distance between the laboratories and, in the case of interdisciplinary work, along with the disciplinary distance between the laboratories. The best way to encourage trustworthy behavior is to create an intellectually open research environment, one that accommodates the ambiguity of everyday practice.

When editors and reviewers evaluate research manuscripts, they also have to trust that the work submitted actually was carried out as described and that it produced the results reported. The role of editors and reviewers is not to replicate and validate the research. Sometimes they raise concerns about statistical analyses or computer manipulation of data images. Overall, however, the focus of initial review is whether the work appears worthwhile to publish based on quality of the data, reasonableness of interpretation, and relevance. Only later, after discovery claims become public through publication, does the process of replication and validation begin through which discovery claims come to be seen as credible or flawed.

Betrayal of trust

Sociologist Robert Merton attributed "the virtual absence of fraud in the annals of science" to the norm of disinterestedness. Often misunderstood, Merton did not mean that scientists are naturally inclined to be disinterested or objective. Rather, he meant that the robustness of the credibility process provides a strong incentive for investigators to behave in a trustworthy fashion. "Involving as it does the verifiability of results, scientific research is under the exacting scrutiny of fellow experts. Otherwise put...the activities of scientists are subject to rigorous policing, to a degree perhaps unparalleled in any other field of activity" (36).

Nevertheless, betrayal of trust occurs. Soon after I joined Paul Srere's laboratory as a postdoctoral fellow, he told me the following story. As a graduate student, Srere said, he had made a breakthrough discovery that was "stolen" by another biochemist who had learned about the new finding while visiting Srere's Ph.D. thesis adviser. By stolen, Srere meant that the other biochemist subsequently pursued the discovery in his own laboratory and was able to publish the findings before Srere. The other biochemist got the credit. Even after this event occurred, Srere told me that he continued to believe in discussing unpublished results, even though others might take advantage of the information: "If you only have one good idea in your career, then it will not matter much if someone steals it. If you have a lot of good ideas, then it will not matter much if someone steals one from time to time."

When I was an assistant professor, I had my own experience of a somewhat related sort, except the outcome was favorable. My collaborators and I submitted a paper describing a unique aspect of protein movement on the outer surfaces of tissue cells. A couple of months later we received reviewers' comments from the journal and a decision letter from the editor declining to publish our manuscript.

A week after that, a postdoctoral fellow working at a major Ivy League medical center called me for more information about our methodology. I asked him how he had heard about our studies. He told me that our work had been presented in a laboratory discussion group that he attended. Since submitted manuscripts are privileged and confidential information, no discussion of our paper should have occurred at a laboratory meeting. We called the journal editor and complained. He decided to manage the breach of confidentiality by publishing our paper and asked us to revise the manuscript and respond to the criticisms that had been raised by the reviewers. Shortly thereafter, we received another decision letter, this one accepting our manuscript.

During the late 1970s, several instances occurred in which scientists flagrantly betrayed the trust of the community by fabrication (making up their research data) and plagiarism (publishing the work of others as if the work were their own). These cases became sufficiently publicized to catch the attention of Congress. Former U.S. vice president Albert Gore (then a congressman) opened the 1981 Congressional hearings on fraud in biomedical research, stating: "At the base of our investment in research lie the trust of the American people and the integrity of the scientific enterprise. If that trust is threatened...then not only are the people placed at potential risk, but the welfare of science itself is undermined" (37).

The scientific community was unprepared to deal with overt scientific misconduct. Most institutions lacked formal guidelines or procedures. Such matters had been dealt with informally in the past. How should *scientific misconduct* be defined?

In 1989, the U.S. PHS provided the first definition for scientific misconduct:

(1) fabrication, falsification, plagiarism, deception or other practices that seriously deviate from those that are commonly

accepted within the scientific community for proposing, conducting or reporting research; or (2) material failure to comply with Federal requirements that uniquely related to the conduct of research. (38)

Controversy ensued. Like the Government Performance and Results Act, the PHS definition appeared to ignore the realities of everyday practice of science. Many researchers as well as scientific professional organizations objected to the phrase "other practices that seriously deviate." As FASEB president Howard Schachman testified:

It is our view that this language is vague and its inclusion could discourage unorthodox, novel, or highly innovative approaches, which in the past have provided the impetus for major advances in science. It hardly needs pointing out that brilliant, creative, pioneering research deviates from that commonly accepted within the scientific community. (39)

The dilemma was how to define misconduct in a way that could accommodate the ambiguity of practice. Consider the research of Nobel Laureate Robert A. Millikan, who selected 58 out of 140 oil drops—the "golden events"—from which he calculated the value of the charge of the electron (40). *Honor in Science*, a booklet published by the international honorary research society Sigma Xi, called Millikan's behavior "one of the best known cases of cooking" (i.e., falsifying data by unrepresentative selection) (41). David Goodstein, when awarded the Sigma Xi McGovern Science and Society Award, used his award lecture to defend Millikan against the charges put forth in *Honor in Science* (42).

The difficulty in assessing Millikan's behavior is that at the edge of discovery, distinguishing data from experimental noise rarely is clear-cut. Although heuristic principles can be helpful, an investigator's experience and intuition—creative insight—often will determine what counts and what does not. In any particular case, the way the results are selected by one investigator might appear arbitrary and self-serving to another—even an example of misconduct. In their 1992 report *Responsible Science*, perhaps the National Academies Panel on Scientific Responsibility and the Conduct of Research was thinking of Millikan when they wrote: "The selective use of research data is another area where the boundary between fabrication and creative insight may not be obvious" (43).

Not only is data selection a common and necessary feature of much research, but also scientific publications frequently rewrite history. Historian Frederic Holmes, in his biographical study of Nobel Laureate Seymour Benzer, commented:

> Like most modern scientific papers, Benzer's "Fine Structure of a Genetic Region in Bacteriophage" is a logical reconstruction of the experiments, observations, and arguments supporting his conclusions. It has little narrative structure and does not purport to follow the investigative pathway from which it came. (44)

If one is looking for an inclusive, historically accurate picture of everyday practice, the only place such a picture might be found is in research notebooks (written, photographic, electronic, etc.). Keeping a complete notebook, I tell graduate students and postdoctoral fellows, is what gives them permission to publish only a representative selection of their data and to do so in a logical rather than historically accurate fashion. Later on, they should be able to defend their choices if asked for an explanation.

Controversy over how to define scientific misconduct was resolved in the United States when the Office of Science and Technology Policy (OSTP) finally replaced the various definitions used by the U.S. PHS and other federal agencies. The new government-wide definition dropped the "other practices that seriously deviate" clause and limited misconduct to "fabrication, falsification, or plagiarism in proposing, performing, or reviewing research, or in reporting research results." In addition, the OSTP definition excluded "honest error or differences of opinion" and established that "a finding of research misconduct requires that there be a significant departure from accepted practices of the relevant research community" (45). The latter aspect of the definition recognizes the heterogeneity of everyday practice across different research communities.

Conflict of interest

In my experience, most scientists consider research misconduct to be pathological and self-destructive behavior, but also rare. Neither public trust in scientists nor the prestige of science as an occupation appears to have declined in recent years, notwithstanding the publicized cases of scientific misconduct (22). What should be far more troublesome to scientists and the public alike is the threat of conflict of interest. Conflict of interest can interfere with every aspect of the practice of science, including analysis and publication of research findings, sharing of research knowledge and tools, and decision making by public advisory committees.

According to the National Academies' definition, a conflict of interest is "any financial or other interest which conflicts with the service of the individual because it (*i*) could significantly impair the individual's objectivity or (*ii*) could create an unfair competitive advantage for any person or organization" (46). Not only soft money support for professorial salaries, but also peer review of grant application and research manuscripts might be considered inherently conflicted according to the above criteria. These conflicts need not undermine practice of science as long as there is coincidence between the best interests of researchers and science as a whole. This coincidence, explained philosopher David Hull, accounts for the high degree of integrity of researchers: "What is good for the individual scientist is actually good for the group. The best thing that a scientist can do for science as a whole is to strive to increase his or her own conceptual inclusive fitness" (47).

Conceptual inclusive fitness means having one's ideas and discoveries become the prevailing beliefs of the scientific community—gaining influence, recognition, and sometimes acclaim for one's work. Because success typically depends on doing important work and getting the right answer, advancing science also advances one's own self-interests. Getting the wrong answer, on the other hand, can have opposite effects—decreased recognition and influence, unfunded grants, fewer trainees, and, ultimately, conceptual extinction! Hull's thinking coincides with Alexander Hamilton's comment that "the best security for the fidelity of mankind is to make their [political leaders'] interest coincide with their duty" (48).

If personal wealth becomes a researcher's overriding interest, then what is good for the individual scientist may no longer be good for science as a whole. Indeed, the potential change in the research environment from "publish or perish" to "patent and prosper" threatens the health of biomedical research (49). This threat arises because the values inherent in "patent and prosper" differ in fundamental ways from those of "publish or perish." Rather than advancing science as a whole, success in entrepreneurship and industry means advancing the interests of the company as determined by the board of directors and shareholders. Privacy replaces public ownership of knowledge. Time-consuming negotiations replace open sharing of resources. Keeping research findings secret or subject to editorial control replaces publishing the results as quickly as possible. Only in an increasingly patent-and-prosper environment would it become necessary for the NIH to remind investigators that "excessive publication delays or requirements for editorial control, approval of publications, or withholding of data all undermine the credibility of research results and are unacceptable" (50). Equity holdings represent the best example of how the relationship between what is good for the individual and what is good for science lose their connection. The value of equity holdings in a company can increase dramatically in response to new research findings. Investigators who divest their equity holdings at the time of the increase can benefit financially even if the new research findings later turn out to be flawed.

Figure 4.3 shows a dramatic example of equity influenced by research. The discovery that embryonic stem cells could be derived from IVF human embryos was reported in the journal *Science* in 1998 (51). The biotechnology company Geron Corporation provided funding for the research. The circles in the figure show the financial impact of the discovery in terms of daily trading of Geron stock for the month of November 1998 and, by contrast, January 2006. On the day that the human embryonic stem cell paper was

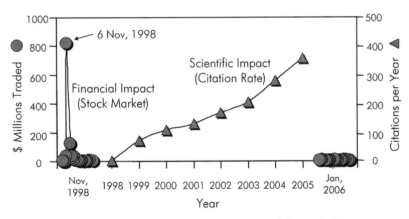

Figure 4.3. Separation of Financial Impact and Scientific Impact

published—November 6, 1998—the price of Geron stock reached almost $25/share. The stock's value had been about $6/share the previous week. More than 40 million shares were traded, 100 times more than usual. A week after the discovery, Geron's stock value was back to around $10/share and the volume was less than 500,000 shares.

The triangles in figure 4.3 show the scientific impact of the discovery as measured by the paper's citation rate. To put these data in perspective, the average paper published in *Science* receives about 15 citations in the year after publication. The human embryonic stem cell paper received almost 100. Moreover, the paper's citation rate continued to increase, reaching almost 400/year in 2005. The scientific impact has been enormous. However, the stock market has not responded further. November 6, 1998, stands out in the financial history of Geron as a day on which a lot of money could have been made even though the credibility of the stem cell discovery claim had yet to be established.

Could a paper important enough to receive 100 citations in the year after publication turn out to be wrong? In 2004, a Korean scientist named Woo Suk Hwang and co-workers published another groundbreaking human stem cell research paper in *Science* (52). In the year after publication, that paper was cited 118 times. A year later, the work turned out to be bogus. Hwang was indicted on charges of fraud, embezzlement, and violation of a Korean bioethics law. Hwang's American collaborator and co-author on a follow-up 2005 paper Gerald Schatten was accused of *research misbehavior* by his university:

> Dr. Schatten's listing as the last author not only conferred considerable credibility to the paper within the international scientific community, but directly benefited Dr. Schatten in numerous ways including enhancement of his scientific

reputation, improved opportunities for additional research funding, enhanced positioning for pending patent applications, and considerable personal financial benefit. (53)

Patent and prosper

For science as a whole, patent and prosper is not a new problem. Although many would argue that financial protection allowed by patents is a necessary factor in industrial development of science and technology, there never has been consensus on this matter. Benjamin Franklin opposed patenting of inventions and believed that achieving the goal of science in the service of society would be accomplished best by making discoveries and inventions freely available to everyone. Commenting in his *Autobiography* on why he declined to patent the Franklin stove and other famous inventions, he wrote: "I declined it from a principle which has ever weighed with me on such occasions, viz., that as we enjoy great advantages from the inventions of others, we should be glad of an opportunity to serve others by any invention of ours, and this we should do freely and generously" (54). Ironically, in the same year as Benjamin Franklin's death (1790), the U.S. Congress passed the Patent Act to establish the U.S. Patent Office. Doing so fulfilled Article I, Section 8, clause 8 of the U.S. Constitution: "To promote the progress of science and useful arts, by securing for limited times to authors and inventors the exclusive right to their respective writings and discoveries."

Along with the financial protection afforded by patents come inevitable patent disputes between scientists ranging from ownership to infringement. The first scene of the exhibit called "Science in American Life," which was on display for many years at the Smithsonian National Museum of American History, showed chemist Ira Remsen and his postdoctoral trainee Constantine Fahlberg in their laboratory at Johns Hopkins University. Remsen founded the

Department of Chemistry at Johns Hopkins University in the late nineteenth century and started the *American Chemical Journal*. The two scientists are shown arguing about credit for their discovery of the artificial sweetener saccharin. Remsen serendipitously made the original observation leading to the discovery, which the two later published together. Fahlberg patented the discovery and claimed to be the senior investigator.

The year 1980 was key for development of the patent-and-prosper environment in biomedical research. Congress passed the Bayh-Dole Act, which encouraged universities to patent technologies that their employees had invented with the help of federal funds. The U.S. Supreme Court decided the case of *Diamond v. Chakrabarty* (447 U.S. 303), after which patents could be issued on forms of life created in the laboratory by genetic engineering. Finally, the first gene-cloning patent was issued to Stanford and the University of California, San Francisco (UCSF). Subsequent licensing of the Stanford/UCSF gene-cloning patents by the biotechnology industry generated well over $100 million in revenues for these two academic institutions. One of the discoverers of gene cloning, Herbert Boyer, started the first biotechnology company—Genentech, Inc.

The three-part combination of patents, licenses, and biotechnology startup companies became a model for how universities might benefit from technology transfer. In 2004, U.S. colleges and universities received more than $1 billion in licensing revenues, executed 4,000 deals for licenses and options, and spun off 425 new startup companies (55). Notwithstanding these accomplishments, most universities still spend more on technology transfer than they earn. Whether overall innovation and competitiveness actually has benefited from Bayh-Dole remains controversial.

Not all the financial news has been good. Scientists in colleges and universities frequently use patented inventions in their work without paying licensing fees—what is called the "experimental use" exclusion. When patent infringement has been claimed, U.S. courts in the past depended on the distinction between commercial and noncommercial research activity to determine if the experimental use exclusion was applicable. This approach changed in the 2002 case of *Madey v. Duke University*. The Federal Circuit Court concluded that

> major research universities, such as Duke, often sanction and fund research projects with arguably no commercial application whatsoever. However, these projects unmistakably further the institution's business objectives.... Regardless of whether a particular institution or entity is engaged in an endeavor for commercial gain, so long as the act is in furtherance of the alleged infringer's legitimate business and is not solely for amusement, to satisfy idle curiosity, or for strictly philosophical inquiry, the act does not qualify for the very narrow and strictly limited experimental use defense. (56)

How the *Madey v. Duke* decision will affect academic researchers in the future and to what extent they will need to acquire licenses to carry out studies that involve patented research tools remain to be seen.

Disclosure or recusal

Because of the large scale and rapid expansion of the patent-and-prosper environment, the question of how to regulate financial conflict of interest in biomedical research has become a major science policy concern. The problem already was highlighted in hearings before the 1990 U.S. Congress that focused in part on conflicts of interest in human clinical research trials. In one example regarding a drug called tissue plasminogen activator (t-PA), the congressional committee concluded that

> NIH ignored information that its grantees owned stock in the company whose product they were studying.... t-PA research and publication decisions with implications for the safety of research subjects were made by scientists with a vested interest in proving the effectiveness of t-PA.... Informed consent forms did not adequately warn potential patients of the possible dangers of participating in the NIH-funded research.... Objective scientific research and Food and Drug Administration (FDA) review for thrombolytic agents have been hampered by consulting and promotional practices. (57)

The committee recommended that Congress should enact legislation if the U.S. PHS did not develop regulations to restrict financial ties for researchers who conduct evaluations of products or treatments in which they have a vested interest.

Within a few years, a new rule was established by the U.S. Department of Health and Human Services (HHS) called "Objectivity in Research":

> to insure that the design, conduct, or reporting of research funded under PHS grants, cooperative agreements or contracts will not be biased by any conflicting financial interest [defined as at least $10,000 or 5% ownership interest in an entity] of those investigators responsible for the research. (58)

The underlying principle of the new objectivity guidelines was that disclosure of potential conflicts of interest would be sufficient to protect the research process. Although research institutions were required to report conflicts of interest to funding agencies, precisely how institutions should manage conflicts was not specified. Moreover, the new regulations did nothing to integrate disclosure of conflicts with the human research protections system. A nationwide survey on implementation concluded that

> there is substantial variation among policies on conflicts of interest at medical schools and other research institutions. This variation, combined with the fact that many scientific journals and funding agencies do not require disclosure of conflicts of interest, suggests that the current standards may not be adequate to maintain a high level of scientific integrity. (59)

The new "objectivity" regulations focused on individual researchers. Nothing was said about the emerging problem of institutional conflict of interest. The American Association of Medical Colleges voiced the concern that "existing institutional processes for resolving competing interests may be insufficient when the institution has a financial interest in the outcome of research and the safety and welfare of human subjects are at stake" (60). In a 2001 report to Congress, the U.S. General Accounting Office recommended that HHS develop specific guidance or regulations concerning institutional financial conflicts of interest (61).

Further compounding the problem of how to manage conflict of interest is what has been called the "funding effect in science." That is, meta-analysis of human research studies has shown a trend that industry-sponsored drug trials tend to reach pro-industry conclusions (62). If a funding effect in science really exists, then no mechanism of disclosure can deal adequately with conflict of interest. Rather, what would be needed is a financial firewall between academic researchers and industrial sponsors—in short, recusal rather than disclosure (63). A form of recusal is precisely what the NIH imposed on its employees in February 2005 (64).

Transition

The biomedical research environment is in transition. How far patent and prosper develops remains to be seen. Recently, my laboratory prepared immortalized human fibroblast cells to use in our research. I thought that the properties of these cells might make them useful as a research tool for others, so I contacted the American Type Culture Collection (ATCC) to see if they wanted the cells for their collection. ATCC sent me back a submission form. I was not sure who should sign the form, so I called our technology transfer office. Instead of telling me who should sign the form, they sent me a multipage intellectual property (IP) questionnaire. I said, "I don't see the point of filling out a complete IP questionnaire to give ATCC a cell line with no commercial value." The technology transfer office told me that they would make that assessment. "In some instances, we may decide it is in the best interest of the university to seek appropriate protection for a new invention (patents, copyright, etc.)." I decided to forget about trying to give the cells to ATCC.

If the technology business model of the research university continues to expand, then peer review eventually will require confidentiality and nondisclosure agreements complete with remedies for noncompliance. Should these changes occur, I believe that the underlying cause will be the research universities and centers and not the researchers themselves. Funding organizations such as the National Science Foundation and the NIH can do a lot to maintain the academic model through their policies regarding sharing of resources, peer review, and so forth. Rather than patent and prosper, what continues to be the primary goal for most scientists whom I know is to have their ideas and discoveries become the prevailing beliefs of the scientific community.

5

INFORMED CONSENT AND RISK

The Intersection of Human Research and Genetics

When it comes to everyday practice of research involving human subjects, the problem of conflict of interest mentioned in chapter 4 is part of a broader set of ethical concerns. *Time* magazine's April 22, 2002, cover shows a woman wearing a hospital gown sitting inside a large cage, with a banner reading "How Medical Testing Has Turned Millions of Us into Human Guinea Pigs." The cover article discusses the dilemma of conflict of interest and then goes on to describe troubles with obtaining informed consent and evaluating research risk (1). I begin this chapter with an overview of human research ethics. I then focus on the increased uncertainties about informed consent and risk for research in the rapidly developing field of genetic medicine.

Midway through the first clinical research study that I directed, a human research ethics dilemma arose. We were carrying out a phase II clinical trial to learn if the biological adhesion protein fibronectin might be used therapeutically to promote healing of chronic leg ulcers. One of the subjects in our study had an ulcer on each leg. We treated one of her ulcers with fibronectin solution and the other with an inactive solution lacking fibronectin. At the end of the three-week experimental period, we thanked her for her participation. She was surprised that we wanted to stop. The ulcer on her left leg was feeling better for the first time in five years. She asked if we could keep on treating her left leg and begin to treat her right leg the same way.

Her request for continued treatment raised all sorts of unexpected issues. One of her ulcers had received solution "A." The other had received solution "B." Neither the research subject nor the research nurse interacting with the subject knew which solution contained the fibronectin and which contained the inactive material. This type of experimental research design is known as *double-blind*. But, I had been naive in coding the treatment samples. Determining which of her ulcers had received the fibronectin preparation would potentially undermine the double-blind experimental design for other subjects. In addition, what if the ulcer responding positively had been receiving the inactive material? What should we do then? Even if it turned out that the healing ulcer had received fibronectin, we did not have any extra fibronectin available. The organization supplying the fibronectin might not want to increase the limited quantities that they had been giving to us. Our study was intended as research, not ongoing therapy. What obligations did we have?

Concerns about how to achieve ethical conduct of research with humans increasingly became a matter for national debate in the United States in the 1960s. The birth defects tragedy caused by giving pregnant women the drug thalidomide highlighted questions about practices of the pharmaceutical industry and physicians with regard to use of inadequately tested medications. At the same time, national surveys revealed a lack of common standards or procedures at medical centers and hospitals for carrying out human research. Decision making regarding ethical and other matters was left primarily to the investigators conducting the research (2). The widespread failure of the status quo became clear in 1966 when an article in the *New England Journal of Medicine* criticized ongoing ethics lapses in the United States at "leading medical schools, university hospitals, private hospitals, governmental military departments,...Veteran's administration hospitals, and industry" (3).

The 1960s also was the decade in which human transplantation medicine achieved success. First came kidney, lung, and liver transplants. The first successful human heart transplant occurred in December 1967. The patient survived for 18 days. Shortly afterward, former Vice President Walter Mondale (then a U.S. senator) called for formation of a Commission on Health, Science, and Society.

> The transplantation of human organs already has raised such serious public questions as who shall live and who shall die. But coming techniques of genetic intervention and behavior control will bring profound moral, legal, ethical, and social questions for a society which one day will have the power to shape the bodies and minds of its citizens. (quoted in 2)

The commission envisioned by Mondale would have studied the legal, social, and ethical implications of medical research. However, most physicians, researchers, and administrators opposed regulatory oversight of human research at the national level. They argued that the research community in the local setting where the studies were occurring was best able to determine what research should be done and how. That opposition mostly evaporated after the now infamous Tuskegee syphilis experiment became public knowledge in 1972.

> For forty years between 1932 and 1972, the U.S. Public Health Service conducted an experiment on 399 black men in the late stages of syphilis. These men, for the most part

illiterate sharecroppers from one of the poorest counties in Alabama, were never told what disease they were suffering from or of its seriousness. Informed that they were being treated for "bad blood," their doctors had no intention of curing them of syphilis at all. The data for the experiment was to be collected from autopsies of the men, and they were thus deliberately left to degenerate under the ravages of tertiary syphilis—which can include tumors, heart disease, paralysis, blindness, insanity, and death.... Even when penicillin—the first real cure for syphilis—was discovered in the 1940s, the Tuskegee men were deliberately denied the medication. (4)

In the aftermath of the Tuskegee revelations, congressional hearings led by Senator Ted Kennedy ultimately led to establishment of the National Commission for the Protection of Human Subjects of Biomedical and Behavioral Research. The commission was given the responsibility to "identify the basic ethical principles that should underlie the conduct of biomedical and behavioral research involving human subjects and to develop guidelines which should be followed to assure that such research is conducted in accordance with those principles" (5).

Principles of human research ethics

Historically, philosophers often try to identify basic ethical principles by focusing on different aspects of human behavior: individuals as actors, the actions that they undertake, or the consequences of their actions. Some theories of ethics emphasize the moral character of the actors. Others emphasize the specifics of the actions carried out. Actions can be viewed as right or wrong intrinsically. A third approach emphasizes the consequences of an action for the actor and for others (6).

By the time the U.S. National Commission began its work, a world consensus had developed that how human research was carried out—especially the manner in which the human research subject was treated—was more important than the moral character of the researcher or the potentially beneficial consequences of the research for science and society. The 1949 Nuremberg Code began with the statement, "The voluntary consent of the human subject is absolutely essential." The World Declaration of Helsinki (first adopted in 1964) stated, "Concern for the interests of the subject must always prevail over the interests of science and society" (7).

The U.S. National Commission's conclusions—known as the *Belmont Report*—established three basic ethical principles of human research: *respect for persons, beneficence,* and *justice* (5). Respect for persons means treating human research subjects as independent and self-directing individuals. For someone to become a subject in a particular research study requires the person's informed consent. The individual should understand the nature of the research and its risks and possible benefits, and should choose to participate voluntarily. The second principle, beneficence, means designing experiments and treating human subjects to minimize possible risks of the research while maximizing possible benefits. The third principle, justice, means fair recruitment and selection of human research subjects—those who bear the burdens of research should be equally likely to share the benefits.

The Belmont Report became the foundation for subsequent U.S. human research protection and oversight regulations. At the center of the system, institutional review boards (IRBs) were authorized and established. IRBs were to function as local review committees at hospitals, research centers, and so forth. They became responsible for ensuring that every aspect of human research from experimental design to informed consent meets national standards.

That the Nuremberg Code and World Declaration of Helsinki focused on well-being of research subjects reflected a reaction to experiments carried out by Nazi physicians on concentration camp prisoners during World War II. An international court later condemned these experiments as "crimes against humanity." In addition, the U.S. National Commission was reacting to questionable human research practices in the United States itself. In discussing the principle of justice, the commission explicitly recognized the historical problem of human exploitation for the purposes of advancing research. They mentioned not only the Nazi experience, but also the use of disadvantaged, rural black men for the Tuskegee syphilis study.

Ethical challenges

Despite development of extensive human research protections programs in the United States and elsewhere, significant problems continue to challenge the ethics of practice. In part, the problem is that, by its very nature, research on humans presents something of an ethical paradox. Although the interests of the subject should prevail over those of science and society, inevitably, the *risks* of research most likely will affect the subject, whereas the *benefits* most likely will affect persons in the future and thereby advance the interests of science and society. For instance, in the United States, fewer than 1 in 10 new drugs tested on human research subjects turn out to be sufficiently safe and effective to be approved for marketing (8).

Another ethical challenge concerns the use of *randomized, placebo-controlled* experimental design. The features that make each person unique and special from a humanistic point of view can confound scientific research. Rather than uniqueness, research seeks to study typical examples of diseases with the goal of developing therapies to treat anyone (or anyone of a definable subset of persons) who happens to have a disease. Randomized, placebo-controlled experimental design accomplishes this goal by assigning research subjects arbitrarily, such as by a coin toss, to receive either the active therapy or a treatment expected to be completely inactive, the *placebo*. In our study on chronic ulcers, the inactive solution lacking fibronectin was the placebo.

Double-blinded, randomized, placebo-controlled trials often are described as the gold standard of clinical research. This experimental design is believed to be "as free of assumptions and reliance on external information as it is possible to be" and to offer important advantages over other research designs (9):

- Minimizing the effects of expectations on the part of subjects (i.e., the placebo effect)

- Minimizing the effects of expectations on the part of researchers

- Providing an internal control (i.e., the placebo), which increases the sensitivity of the research to detect differences in effectiveness and safety

- Decreasing the total number of subjects required to obtain reliable information

Notwithstanding the benefits, randomized, placebo-controlled research design often has been described as turning humans into guinea pigs, especially if any therapeutic interventions already have been shown to be helpful in treating the disease under study. For this reason, considerable criticism has been directed toward research in third-world countries where, with the availability of other treatments lacking, placebo-controlled research design sometimes is carried out even though the research would not have been permitted in the countries sponsoring the work (10).

Perhaps the greatest challenge to human research ethics is adequacy (or inadequacy) of the informed consent process. Prospective subjects frequently cannot distinguish research from therapy. In our study using fibronectin to treat chronic ulcers, the situation of the patient requesting continued treatment reflects the difficulty of making this distinction. Until the moment when we told her "thank you very much" and "the experiment is over," there was little in our interaction that could have allowed her to understand clearly the difference between being the subject of a research experiment versus a patient getting therapy (11). She had a disease that required treatment. She was receiving treatment in a health care provider setting. True, at the beginning of the informed consent process, the health care professional made the offer, "You are invited to participate in this research because..." But what difference did it make that the goal of the research was to test an experimental therapy that might not be successful? What difference did it make that she might get the placebo instead of the active drug? This patient had tried all sorts of therapies over the previous five-year period. They all had failed her even though many already were approved for clinical use. What did she have to lose? The experimental therapy was under investigation precisely because the new treatment might be beneficial. Otherwise, what would have been the point of doing the research?

The idea of informed consent has two connotations. The first, which is easier to document in practice, focuses on the prospective research subject's competence to make choices and to authorize participation in research. The second sets a higher standard by focusing on the prospective subject's intentional decision-making capacity. Can the person truly understand what participating in a particular research study implies and choose to participate based on that understanding (12)? If a health care professional says "research," but the person (and perhaps the health care professional, as well) thinks/hopes/hears "therapy," then instead of informed consent, one has what is called *therapeutic misconception* (13).

For prospective subjects to avoid therapeutic misconception and understand the difference between research and therapy, one imagines that very blunt language focusing on research endings might be necessary—something like the following:

> You are invited to participate in a research study. Let me make clear what research means. If we were treating you as a patient and the experimental therapy helped your condition, then we would continue to treat you as long as your condition improved. But this is research. For logistic/financial/legal reasons, the manufacturer/developer of the experimental therapy will not permit us to continue treatment beyond the experimental test period. Therefore, at the end of the experiment we will stop your treatment even if your health has been improving and even if your condition might deteriorate because of discontinued treatment. In the highly unlikely event that the experimental treatment results in your death, even this tragedy would provide us with valuable data.

Of course, this language is intentionally provocative and overstated. Research subjects sometimes can continue experimental therapies at the end of research protocols if their condition is improving (i.e., so-called compassionate care). Such was the case for our patient, who received fibronectin therapy on both legs for an additional six months. The point I am trying to emphasize is that the informed consent interaction needs to jolt a prospective subject out of what philosophers call the *natural attitude*—everyday life experience taken for granted in an unexamined manner (14). "Wait a moment! You want me to do what?" A person who does not experience this discontinuity likely will become a research subject without realizing what becoming a research subject means.

Unfortunately, rather than becoming more blunt, the language of human research appears to be going in the opposite direction by becoming more "sensitive." The euphemistic expression *research participant* has been suggested as a replacement for research subject so as not to hurt anyone's feelings or imply that people are being treated like guinea pigs (15). Unlike patients, research subjects rarely if ever play any authentic participatory role in choosing and modulating their treatment. Consequently, calling subjects participants has the potential to mislead further everyone involved with the informed consent process and to blur rather than sharpen the distinction between research and therapy.

Risk/benefit analysis also is problematic for informed consent. Assessing potential risks and benefits is necessary, according to the Belmont Report, for investigators to decide if their research is properly designed, for review committees to decide if the risks of research are justified, and for prospective subjects to decide whether to participate in the research. What kind of blunt language might be required to help potential research subjects appreciate risk/benefit uncertainty?

> Adverse drug reactions are possible, but we cannot be sure. Other negative consequences are possible since we are ignorant of the full range of potential outcomes. One of the researchers might make a mistake in carrying out the research. Of course, whether the new drug/treatment will improve your condition is unknown as well. (adapted from 16)

"Wait a moment! You want me to do what?"

Genetics research

Modern genetics research exacerbates the problem of informed consent. In genetics research, even the question of "who" should be considered a research subject becomes a matter for debate. This transformation has come about because genetic medicine changes the meaning of words such as "disease," "patient," and "treatment."

By the late 1960s, the implications of modern genetics for human well-being had become a matter of general public recognition and concern. Genetic enhancement was one of the items on Mondale's list when he proposed formation of the Commission on Health, Science, and Society. The April 19, 1971, cover of *Time* magazine shows a man and woman, side by side, each embedded in a deoxyribonucleic acid (DNA) double helix. The banner reads: "The New Genetics: Man into Superman."

> Man's molecular manipulations need hardly be confined to the prevention and cure of disease. His understanding of the mechanisms of life opens the door to genetic engineering and control of the very process of evolution. DNA can now be created in the laboratory. Soon, man will be able to create man—and even superman. (17)

The *Time* issue was remarkably visionary in its anticipation of what was to become the contemporary human genetics research agenda.

The reader need not understand the intricacies of modern genetics to understand the discussion that follows. Know, however, that the field of genetics continues to develop remarkably fast. Even the meaning of the word *gene* is in flux. Early in the twentieth century, genes referred to characteristics of animals and plants that could be inherited from parent to offspring, and whose transmission occurred according to specific mathematical relationships. Research beginning around the time of World War II identified the chemical nature of genes and led to the idea of a one-to-one correspondence between genetic sequences and proteins. As science students in the 1960s, we visualized the arrangement of genes to be something like beads on a string and learned the relationship described by the then commonplace expression: *one gene—one enzyme*. Increasingly, however, beads and string appear interchangeable. Complex regulatory mechanisms permit genetic mixing and matching to occur in multiple ways. Now, a consensus definition of the meaning of a gene includes not only the earlier notion of inherited characteristics, but also all of the more recently discovered regulatory features: A gene is a "locatable region of genomic sequence, corresponding to a unit of inheritance, which is associated with regulatory regions, transcribed regions and/or other functional sequence regions" (18).

The above definition probably also will turn out to be an oversimplification. Only 1–2% of the human genome actually codes for proteins. Much of the remainder is transcribed into non-protein-coding ribonucleic acids—so-called *genetic dark matter*—whose regulatory functions and evolutionary significance are just beginning to be understood (19).

What is a genetic disease?

The impact of genetics on the meaning of disease continues a transition already under way from understanding disease as a concrete condition of the present to a probabilistic condition of the future. If a person suffers symptoms of pain or loss of function, she might consult a health care provider hoping to correct the problem. In response, the health care provider will try to identify and treat the cause of the symptoms. Traumatic injury represents a clear-cut example. A person falls, experiences pain and weakness in one leg, and has difficulty standing. The individual goes to a hospital emergency department, where the problem is diagnosed and treated as a broken hip. In this scenario, the person is the first to know, "Something is wrong."

Advances in medical diagnostic technology increasingly permit illness-associated physiology or anatomy to be identified before the patient experiences any "dis-ease" symptoms. As a result, the health care provider sometimes knows before the patient that something might be wrong. Typical examples are elevated blood pressure (hypertension) and decreased bone density (osteoporosis). In the case of elevated blood pressure, the physician might warn the patient about possible cardiovascular consequences, including heart attack and stroke. In the case of decreased bone density, the physical might tell the patient, "If you fall, there is a much greater risk that you will fracture your hip compared to a person with normal bone density." Now we use the word *disease* in two different ways: how a person feels (i.e., disease as "dis-ease") and what the health care community calls the person's condition (i.e., disease as a named illness). The modern patient can arrive at a health care provider's office for a routine checkup feeling just fine, but leave feeling psychologically not so well, newly aware of the illness.

Genetic medicine takes the separation one giant step further by enabling predictions regarding the person's predisposition to develop disease later in life even before physiological or anatomical disease changes have begun to occur. To understand the idea of "predisposition" requires explaining the role of genetic mutation in disease development. Many common illnesses, including cancers, heart disease, diabetes, and osteoporosis, exhibit a complex interrelationship between environmental factors and genetic mutations. These illnesses are referred to as *multifactorial* diseases. Cancers are caused by the accumulation of mutations in several genes that regulate cell growth, death, DNA repair, and migration of specific cell populations. Once mutated, the genes no longer function as before, and as a result, the mutated cells can grow in an unregulated fashion and invade surrounding tissues, sometimes traveling to distant sites in the body (metastasis). Potential cancer-causing mutations occur spontaneously in every person at a low rate. Since several changes have to accumulate before actual disease occurs, people usually develop cancers only after they get older. Many people do not develop detectable cancers before they die from other causes. Exposure to environmental mutagenic agents such as nuclear radiation, tobacco smoke, or certain pesticides and other chemicals will result in a greater rate of spontaneous mutation. Those who have been exposed have a greater likelihood of developing cancers compared to those who by choice or by chance have avoided the environmental risk factors.

Of particular interest to the practice of genetic medicine are the disease-associated gene mutations that can be inherited from one's parents. Because developing a particular type of cancer depends on the mutation of several genes, a person who inherits one of these genes already mutated will require fewer additional spontaneous mutations before developing cancer. As a result, the person's susceptibility will be increased compared to someone without the gene mutation. Breast cancer is the most common type of cancer experienced by women. About 15% of women will develop the disease sometime during their lifetimes. In a small subset of families, some women inherit mutations in genes called *BRCA-1* (breast cancer-1) and *BRCA-2*. Having one of these mutations increases from 15% to 40–90% the likelihood that the woman will develop the disease, but with time of onset influenced considerably by unknown environmental factors (20).

Some inherited genetic diseases can be caused by a mutation in a single gene. A subset of patients who develop osteoporosis have a single-gene disease called osteogenesis imperfecta (OI). In OI, inherited genetic mutations result in decreased synthesis or altered structure of type I collagen, the most abundant protein of bone. All individuals with OI gene mutations eventually develop disease symptoms. Severity and timing vary widely, ranging from crippling at birth to undetectable until teenage or adult years (21). Other well-known single-gene disorders include sickle cell disease, cystic fibrosis, and Huntington's disease. For all of these illnesses, the timing of disease onset and progression can vary a great deal from person to person.

Besides inherited mutations, some genetic diseases are associated with spontaneous changes in the overall organization of genetic material. These changes frequently result in abnormal appearance of chromosomes. The best-known example of chromosomal disease is trisomy 21 (Down syndrome), which gets its name because affected individuals have an extra chromosome 21. Here again, the severity of disease symptoms covers a wide range of possibilities (22).

Identification of disease-associated gene mutations or chromosomal abnormalities is an important step in understanding the underlying causes of disease. Once the mutations are known, they can then be tested for in any individual. Having this information allows predictions to be made about a person's future disease risk before physiologic or anatomic disease-associated changes appear—*presymptomatic* genetic testing.

Although genetic inheritance will be only one factor that influences potential outcomes for an individual's life, many people tend to equate genetics with an individual's future. This way of thinking, *genetic essentialism*, is the central belief of a late-nineteenth-century social program called *eugenics* (good genes). The goal of eugenics was to improve the mental and physical racial qualities of future generations. Eugenics changes the focus of medicine from making people better to making better people (23).

In the United States, eugenics was centered at Cold Spring Harbor Laboratories (CSHL), now a prestigious teaching and research institution. In the early 1900s, the Eugenic Record Office at CSHL exerted considerable influence not only by introducing genetics into U.S. immigration policy, but also by advocating policies of forced sterilization and restrictive mating to improve the human race (24). Many states adopted sterilization policies. When a challenge to Virginia's 1924 Eugenical Sterilization Act reached the U.S. Supreme Court in 1927, the court sided with the state: "It is better for all the world, if instead of waiting to execute degenerate offspring for crime, or to let them starve for their imbecility, society can prevent those who are manifestly unfit from continuing their kind" (25). Now, government-enforced eugenics has been discredited completely. Yet the tendency to equate genetic predispositions with the potential for a healthy and successful life continues.

Given this introduction to genetic disease, how might one inform prospective subjects about the risks of participating in a research study aimed at identifying inherited, disease-associated gene mutations? The informed consent statement used by University of Texas Southwestern Medical School tells prospective subjects: "If the results of DNA tests show that you or anybody else in your family may develop [insert name of disorder], you and other family members could experience serious stress after receiving such information" (26). Onset of genetic disease will occur in the future if onset occurs at all, but other kinds of dis-ease—for instance, psychological—may appear immediately. You might learn something about yourself that you wish you did not know. In addition, depending on the results of your DNA tests, your fiancée might decide to break off the engagement. The judge might award custody of your child to your estranged and soon-to-be-divorced spouse. Sometime in the future you might get a phone call informing you that one of your genetic sequences puts you at very high risk of developing a disease totally unrelated to the original research study in which you agreed to participate.

Who is the genetic patient?

Just as the meaning of disease has been changing in modern medicine, the conventional view of patients and their bodies has been changing, as well. Medical specialization turns individuals into *virtually fragmented bodies* when patients consult different specialists for different body parts—internist, dermatologist, cardiologist, orthopedic surgeon, podiatrist, and so forth. Transplantation medicine turns patients into *mosaic bodies* that have received fragments— for example, kidney, heart, liver, or lung—from other individuals. Ironically, the mosaic body is a body at war with itself because the immunological self sees the incorporated fragments as foreign—"not mine"—and tries to reject them. Most dramatic of all, artificial life support technology permits patients to exist as *bodies with minimal or no brain function.*

Notwithstanding their differences, fragmented, mosaic, and brain-functionless patients each have in common the traditional relationship of one body = one patient. Genetics introduces two new relationships. At one end of the scale, one cell (the embryo) = one patient; at the other end, many bodies = one patient.

Embryonic life represents undifferentiated genetic potential—a time when gene expression integrated with life experience is beginning to give rise to the increasingly historical person. Potentially, the embryo would be ideal for any sort of genetic intervention, therapeutic or preventative. Early embryos at the six- to eight-cell stage can be evaluated for possible genetic abnormalities. The procedure will not harm the embryo's capacity to develop (27). After parents learn the genetics and chromosomal organization of an in vitro fertilized embryo, they can then choose whether to use the embryo for assisted reproduction. Preimplantation genetic diagnosis introduces the possibility of identifying whether one has a "diseased" genome or chromosomes even before anything recognizable as a human body becomes visible.

Developing and perfecting such techniques as preimplantation genetic diagnosis require research with human embryos. Using human embryos for research has been highly controversial, because the research results in embryo destruction. As discussed in chapter 4, opinions about embryo destruction depend a great deal on attitudes toward human personhood. Those who think of embryos as organized-human-tissue-but-not-yet-persons often favor research with embryos, whereas those who think of embryos as persons-not-yet-born tend to be in opposition.

In addition to the problem of personhood, the question of how to define embryo death has become a potential research issue. Experiments carried out on embryos that "die" naturally as a result of development arrest caused by malfunction of some genetic program or adverse external conditions (28) might be acceptable to groups that are opposed to research that causes embryo death. Richard Doerflinger, one of the leading spokespersons for the Catholic Church's position, wrote: "Catholic teaching requires that the corpses of human embryos and fetuses...be respected just as the remains of other human beings. Use of such material [e.g., for research] is not rejected in principle, but must meet moral requirements" (29). Therefore, the ethical considerations and regulatory strategies that authorize research using cells and tissues derived from naturally aborted fetuses might also be applicable to naturally dead embryos. The dead embryo could be a good candidate for some types of investigations although its reproductive potential has been lost.

At the opposite extreme from the one cell = one patient relationship, genetics also gives rise to many bodies = one patient. A good description of the many bodies/one patient relationship can be found in the American Society of Human Genetics statement on "Professional Disclosure of Familial Genetic Information":

> It is clear that genetic information is both individual and familial. This raises conflicts between the duty of confidentiality and the duty to warn....At the very least, it is clear that a health care professional has a positive duty to inform a patient about the potential genetic risks to his/her relatives. Then, depending on the circumstances, the health care professional may have a duty to warn at risk relatives where: the

harm is serious, imminent and likely; prevention or treatment is available; and where a health care professional in like circumstances would disclose. (30)

Stated otherwise, when a health care professional examines an individual patient for inherited genetic disease, the professional also is examining the patient's relatives. Whatever the health care professional learns about the patient will have direct application to family members (the many-bodies patient). The risks of inherited genetic disease about which the health care professional learns might be harmful enough to require warning the many-bodies patient of the danger even if the original individual patient objects. Under some circumstances, U.S. federal regulations regarding privacy permit information to be shared "to prevent or lessen a serious and imminent threat to a person or the public, when such disclosure is made to someone they believe can prevent or lessen the threat (including the target of the threat)" (31).

The same reasoning that establishes the many-bodies patient in genetic medicine also establishes secondary subjects in human genetics research. The potential problem of secondary subjects is an issue that has been of longstanding interest in the social sciences, especially with regard to survey work. In response to a question regarding whether researchers can ask for personal details about other family members, the American Association for Public Opinion Research offers the following guidance:

It depends. This question deals with an issue known as third party consent. Although the Federal regulations do not deal with this issue explicitly, the question has received increasing discussion over the last several years. Many IRBs would conclude that, when an investigator conducting research obtains identifiable private information about a living individual, that individual becomes a research subject, regardless of whether that person is the individual with whom the investigator is having an interaction. (32)

The "increasing discussion" referred to above began after a father complained about a twin study being conducted in 1999 at Virginia Commonwealth University. In later congressional testimony, the father commented that he was surprised that so many of the questions in the survey sent to his daughter concerned himself and other family members. He was shocked by their bizarre nature. He pointed out that "nowhere in the study packet were the words 'informed consent' ever mentioned" (33). Investigation of the complaint by the U.S. Office for Protection from Research Risks (now called the Office for Human Research Protections) led to criticism of the university's IRB for failing to consider the "potential social, psychological, and legal risks" to family members by collecting their detailed medical and social information without consent.

In survey research, information collected about secondary subjects has to be specifically requested. What the researcher learns are beliefs about the secondary subject that may or may not be correct. In genetics research, by contrast, information about secondary subjects comes without asking and offers specific insights into genetic profiles ranging from identical (twins) to more distant, for example, parents, children, and siblings.

Expert panels at the U.S. National Institutes of Health (NIH) and within the Department of Health and Human Services have discussed whether genetics research requires special consideration with regard to secondary subjects (34). As of 2008, no specific recommendations have been put forth to provide guidance on this matter. Currently, local IRBs decide whether secondary subjects count as research subjects and whether the risks involved require informed consent. If genetics research did require informed consent of second-ary subjects, then that requirement could make "a large proportion of current research in human genetics impracticable—with highly detrimental consequences to ultimate public benefit," commented the former leader of NIH's genome research program (35). The American Society of Human Genetics reached similar conclusions when the society advised its membership:

> If, in large family studies, each family member must be enrolled as a "human subject," with informed consent pro-cedures, before any medical information about them can be collected, obtaining family histories will be enormously cumbersome and prohibitive, and will seriously impede medical research. Unless a "waiver" is granted, the project may no longer be feasible. (36)

The problem of secondary subjects exists for groups as well as individuals. The goal of the National Geographic Society's *Genographic Project* is to map the history of human migration world-wide beginning from our common African roots. The project went on hold temporarily in December 2006 when the Alaska Native Medical Center IRB decided that the project did not adequately inform Alaska Native peoples of the risks of the project to the wel-fare of tribal groups as a whole (37).

How are genetic diseases treated?

In preceding sections, I discussed several ways in which medi-cal genetics introduces new ways of thinking about diseases and patients. Now I turn to the meaning of treatment. The American Board of Medical Specialties oversees certification of physicians in the United States. One of the 24 board-certified specialties is called *Medical Genetics*. A medical geneticist is

> a specialist trained in diagnostic and therapeutic procedures for patients with genetically linked diseases. This specialist uses modern cytogenetic, radiologic and biochemical testing to assist in specialized genetic counseling, implement needed therapeutic interventions and provide prevention through prenatal diagnosis. (38)

The inherited genetic disease phenylketonuria (PKU) is an ideal example of how testing for genetic disease makes therapeutic intervention possible. Individuals with PKU have a defective gene for the enzyme that metabolizes the amino acid phenylalanine. During embryonic and fetal life, the mother's normal metabolism will keep the developing child's levels of phenylalanine at normal levels. After birth, newborns will experience severe and irreversible mental retardation as toxic levels of phenylalanine accumulate in their blood. Beginning in the 1970s, widespread use of biochemical screening to detect PKU at birth essentially eliminated this cause of mental retardation because the affected infants could be placed immediately on a low-phenylalanine diet (39). Although inconvenient, continuing on the dietary restrictions allows PKU patients to avoid onset of neurological symptoms throughout their lives.

Unlike PKU, many, perhaps most, inherited genetic diseases do not yet have effective therapeutic interventions. Tay-Sachs disease is an example at the other extreme. Infants with Tay-Sachs usually appear normal at birth, but disease symptoms begin to emerge after a few months. Most infants "become blind, deaf, and unable to swallow. Muscles begin to atrophy and paralysis sets in.... Even with the best of care, children with Tay-Sachs disease usually die by age 4, from recurring infections" (40).

Gene therapy, if possible, would offer the ideal therapeutic intervention to cure genetic diseases. If the mutated genetic sequences responsible for PKU or Tay-Sachs could be corrected or if the mutated genes could be excised and replaced by normal genes, then a cure would have been accomplished at the genome level. Hopeful of such an outcome, clinical research trials involving human gene therapy began in the United States in the 1980s. The field has not developed as expected. Despite the original intent to treat single-gene inherited diseases, by 2008 fewer than 10% of the 1,347 clinical trials ongoing worldwide targeted single-gene defects. Almost 70% were directed toward treating different types of cancers (41). In addition, none of the small number of clinical trials that have advanced to phase III are targeting single-gene diseases. Phase III represents a stage of research when a therapy shown previously to be successful with a small group of patients at one test site is tried with larger groups of patients at several different test sites.

Why has gene therapy for inherited single-gene diseases not progressed further? One reason has to do with evolutionary uncertainty. Unlike conventional engineering in which things are constructed from the ground up with minimal variables, genetic engineering represents intervention in a preexisting biological system whose origins and functions are incompletely understood (42).

A second reason has to do with the subjects of the research. Embryos or fetuses would be the best candidates from a genetic/medical perspective. The longer one waits to intervene, the more widespread will be the effects of a mutation as the individual develops. Greater possibilities will arise for irreversible damage to be caused by mutated genes that fail to carry out normal and necessary functions during development. Nevertheless, from an ethical/policy perspective, embryos and fetuses are the most controversial candidates for gene therapy because of the unpredictability of outcomes and the potential for transmission of these unpredictable outcomes to the next generation. Therefore, at least initially, newborn infants were determined to be the youngest potential subjects of gene therapy research (43).

The third reason that gene therapy has not progressed as hoped is technological. Gene therapy in the sense of repairing or replacing a defective gene is not currently possible. Instead, gene therapy is being accomplished by gene addition—giving individuals a new functional gene. Precisely where the new functional gene will insert in preexisting DNA is not yet controllable. The insertion site can create new and unexpected problems. Such was the case in the outcome of gene therapy clinical trials aimed at treating children with "bubble baby disease" (severe combined immunodeficiency). Because these children have adenosine deaminase (ADA) deficiency, their immune system does not function, and they cannot survive exposure to pathogens found in the normal environment. Of 17 newborns successfully treated by gene therapy—receiving a functional ADA gene—three developed unexpected leukemias because of the way that the added gene inserted into the DNA of their cells (44).

The more successful use of gene therapy for treating cancers and some other diseases compared to single-gene diseases reflects a different underlying strategy—treatment with genes rather than treatment of genes. Rather than aiming to correct or replace disease-associated genes, cancer gene therapy adds new genes that change the cancerous cells and make them more susceptible to being killed—a kind of perverse genetic enhancement.

Although conventional therapy and gene therapy cannot be used at present for treating most inherited genetic diseases, concerned parents still have the prevention option. Disease prevention traditionally refers to public health and population-based analyses of societal and behavioral factors that influence health and disease. These analyses make possible interventions to encourage or eliminate relevant factors—"Stop smoking and exercise more!" Rather than prevention of the disease, prevention through genetic testing ultimately means prevention of individuals who might develop the disease. Such prevention can be accomplished indirectly when adults make decisions regarding whom to marry or whether to have children. Prevention also can be accomplished directly by using preimplantation genetic diagnosis to decide whether to implant an embryo, or by using prenatal genetic testing to decide whether to abort an at-risk fetus.

The consequences of some inherited genetic diseases such as Tay-Sachs are so catastrophic and the biological impairments so completely disabling that many parents might find the prevention option attractive. Deciding what to do becomes much more difficult if an embryo or fetus is destined to develop disease with a less catastrophic outcome. Even if a genetic disease cannot be cured, the consequences of biological impairments often can be reduced when society provides enabling socioeconomic supports and accessible physical environments that allow individuals to lead full lives as much as possible despite their impairments. The absence of such supports encourages selection of the preventive option, sometimes called *backdoor eugenics.* As pointed out by the U.S. Office of Technology Assessment in their analysis of the Human Genome Project, "new technologies for identifying traits and altering genes make it possible for eugenic goals to be achieved through [self-selecting] technological as opposed to [government imposed] social control" (45).

Genetics research and vulnerability

Uncertainty about disease risks, uncertainty about who is an appropriate subject for genetic research, uncertainty about which subjects require informed consent, uncertainty about the eugenic implications of genetics research—taken together, these uncertainties have significant potential to compromise further an already troubled informed consent process.

"Significant doubt," concluded an Institute of Medicine panel,

> exists regarding the capacity of the current [human research protections] system to meet its core objectives. Although all stakeholders agree that participant protection must be of paramount concern in every aspect of the research process, a variety of faults and problems in the present system have been noted. The common finding is that dissatisfaction with the current system is widespread. (46)

The Institute of Medicine report did not specifically mention genetics research as a challenge for human research ethics, but the work of the panel was motivated in part by the death in 1999 of an 18-year-old boy named Jesse Gelsinger who participated in a gene therapy study at the University of Pennsylvania (47).

The Belmont Report introduced the idea that potential subjects are *vulnerable* if they have "compromised capacity for informed consent" (5). Originally, the idea of vulnerability was directed toward groups of individuals that were at increased risk:

> Certain groups, such as racial minorities, the economically disadvantaged, the very sick, and the institutionalized may continually be sought as research subjects, owing to their ready availability in settings where research is conducted. Given their dependent status and their frequently compromised capacity for free consent, they should be protected against the danger of being involved in research solely for administrative convenience, or because they are easy to manipulate as a result of their illness or socioeconomic condition. (5)

More recently, the meaning of vulnerability has been broadened to include anyone in circumstances in which the possibility of obtaining informed consent is compromised. These circumstances can range from insufficient cognitive capacity on the part of prospective subjects to insufficient infrastructural capacity inherent in the research setting (48). Given the increased uncertainties regarding informed consent and risk/benefit analysis, perhaps every potential subject of human genetics research should be considered vulnerable (49).

If vulnerable subjects are entitled to additional protections, then what sort of additional protections might be offered to human subjects of genetic research? Two ideas that have previously been recommended by national advisory committees (for all of human research) seem to me to be particularly appropriate. One concerns an increased public role in decision making. The other establishes compensation for research-related injuries.

An Institute of Medicine panel studying "Scientific Opportunities and Public Needs" concluded that biomedical science requires broader public participation (50). This conclusion led to establishment of the NIH Director's Council of Public Representatives and increased public representation in NIH working groups concerned with diverse issues, including human research. Nevertheless, at the local level of IRBs, public participation continues to be very limited. Federal regulations require only that there be one IRB member whose primary concerns are in nonscientific areas and one member unaffiliated with the institution (and they can be the same person) (51). The National Bioethics Advisory Committee recommended that one-quarter of the membership of IRBs should be members who represent "the perspectives of participants, members who are unaffiliated with the institution, and members whose primary concerns are in nonscientific areas" (52).

The National Bioethics Advisory Committee also discussed ethical considerations underlying the need to compensate research subjects for research related injuries:

First, research participants are entitled to be left no worse off than they would have been had they not participated in the research. Second, without compensation, those with limited access to health care services may not be able to afford treatment and rehabilitation for research-related injuries. Third, even if consent forms and the disclosures within them are comprehensive, completely unforeseen iatrogenic harms [caused by the treatment] might occur in research. Fourth, research participation is often an act of social beneficence in which people agree to volunteer in order to further larger societal goals. Thus, injuries should not be summarily dismissed merely because the individuals were provided information in the consent process and agreed to accept the risk of certain harms. (52)

Ironically, rather than encouraging patient compensation, the only guidance related to compensation for injury that is available on the U.S. Office for Human Research Protections Web site specifies acceptable language for informing prospective subjects that *no compensation will be available*:

- This hospital is not able to offer financial compensation nor to absorb the costs of medical treatment should you be injured as a result of participating in this research.

- This hospital makes no commitment to provide free medical care or payment for any unfavorable outcomes resulting from participation in this research. Medical services will be offered at the usual charge. (53)

Absence of universal health care and fears about projected costs of compensation—how much, how long, who pays—continue to impede any changes in the status quo. As a result, compensation for research-related injuries remains unavailable at most U.S. institutions (54).

The goal of obtaining "truly" informed consent may represent an unreachable ideal for human genetics research or, indeed, for any human research. If this is the case, then increased public participation in decision making and establishing compensation for research-related injuries are all the more important because of the additional protections these changes can provide to human subjects at the beginning and the end of every human research study. Admitting the impossibility of obtaining truly informed consent might be a good starting point to begin reevaluation of the human research protection system.

6

FAITH

More Than One Way to Practice the World

The relationship between science and religion is another important aspect of the interaction between science and society. Discussing this relationship will provide further insights into everyday practice of science and help explain why religious interests sometimes attempt to influence the ways in which science is taught and practiced. Sociologist Dorothy Nelkin pointed out four important areas of contemporary life sciences in which the influence of religion has been felt (1):

- Opposition to teaching evolution in public schools, including attempts to introduce creationism or intelligent design into the science curriculum

- Interference with research using embryos and fetuses

- Criticism of genetics research, for example, gene therapy, genetic engineering, and cloning, as threatening the natural order—humans playing God and tampering with God's will

• Concern about patenting and other commercialization practices of biotechnology and genetics that commodify life and reduce its intrinsic value

The desire by biblical fundamentalists to incorporate creationism and intelligent design into the science education curriculum has been rejected consistently by U.S. courts as a violation of separation of church and state (2). On the other hand, the Catholic Church and other religious groups have been successful in encouraging the ban on federal funding for research that harms human embryos (3).

Of course, practice of science from a purely secular perspective also has its own troubling implications. By "truly" secular, I mean philosopher Peter Singer's analysis that starts with "conscious disavowal that members of our own species have, merely because they are members of our species, any distinctive worth or inherent value that puts them above members of other species" (4). Singer's secularism leads to the conclusion that humanness is an irrelevant category for bioethics. Killing a human being may be wrong but not because humans are members of the species *Homo sapiens*. Rather, it is characteristics like rationality, autonomy, and self-consciousness that make a difference. Because infants lack these characteristics, the death of an infant cannot be equated with "normal" adult human beings or, for that matter, other self-conscious beings. This way of thinking would permit infanticide and euthanasia under some circumstances, yet end much of animal research.

Singer's moral skepticism disturbs me because I personally attribute special status to human life. However, such attribution already has religious overtones. For instance, ethical humanists say that "commitment to the supreme worth (or sanctity) of human life" makes their approach to life a religious faith "without imposing ritual obligations or prescribing beliefs about the supernatural" (5).

The connection between science and religion has been characterized in many ways ranging from "completely separated" to "completely integrated" (6). In this chapter, I suggest that science and religion represent distinct human attitudes toward experience based on different types of faith. These two attitudes can be understood as complementary in the sense that physicist Niels Bohr described complementarity for quantum physics (7). According to this idea, science and religion cannot be separated or integrated but rather coexist in the dynamic tension of a holistic framework.

Conflict

The Catholic Church's prosecution of Galileo in the seventeenth century frequently is cited as a classic (some say mythical) example of the conflict between science and religion. I will use this example to introduce the idea of conflicting beliefs of religion and science about "the way things are." I will leave aside questions regarding the use of political power to settle such conflicts, except for one comment: Whenever fundamentalists, religious or otherwise, are empowered by totalitarian political regimes, then science no longer can be taught and practiced freely.

A key element in the confrontation between Galileo and the Catholic Church concerned the movement of the sun. Aristotle had concluded that the earth was stationary and the sun was moving. The astronomer Copernicus had offered a new interpretation: that the earth and other planets move around the sun. Galileo's observational data and conclusions supported Copernicus' idea against Aristotle. But the Catholic Church was unable to abandon Aristotle. Several biblical passages, including Joshua 10:12, "Sun, stand still upon Gibeon, and Moon in the valley of Ajalon," implied that the sun normally was moving and the earth stationary. If Copernicus and Galileo were correct, then perhaps the Church's understanding of the Bible was incorrect, a possibility that the Church was not prepared to contemplate.

It was more than 100 years before Pope Benedict XIV declared Galileo's work free from error in matters of Roman Catholic doctrine by granting an *imprimatur* in 1741. However, not until the occasion of the 100th anniversary of Einstein's birthday in 1979 did the Church under Pope John Paul II begin a serious effort to reexamine the case. John Paul II called for theologians, scientists, and historians to study the Galileo case and to figure out what went wrong so as to end centuries of mistrust engendered by this infamous affair. Doing so, the Pope suggested, could increase the possibility for collaboration between faith and science that would honor both (8). Several years later, the Pontifical Academy concluded that the Church's error in the case of Galileo was reading Scriptures literally. By so doing, the Church had transferred "a question of factual observation into the realm of faith" (9).

As described above, the conflict between religion and science appears to be a disagreement about the way things are. As one of my teachers commented, "If your religion requires six literal days of creation, then it clashes with science." Frequently, however, the conflicting claims of religion and science are much more subtle. One of my favorite examples of this subtlety is the following stanza from Emily Dickinson's poem "A light exists in spring":

> A color stands abroad
> On solitary hills,
> That science cannot overtake,
> But human nature feels. (10)

Is there really something that human nature can feel but that science cannot overtake? How would one know?

"Were one asked to characterize the life of religion in the broadest and most general terms possible," wrote the philosopher William James in *The Varieties of Religious Experience*, "one might say that it consists of the belief that there is an unseen order, and that our supreme good lies in harmoniously adjusting ourselves thereto. This belief and this adjustment are the religious attitude in the soul" (11). Throughout this chapter, I return to James's definition of the religious attitude. If your religion requires six literal days of creation, then it clashes with science, but if your religion teaches that the unseen order of the world has purpose and meaning, then are you at odds with science?

On a summer beach; a rock comes into sight;
two paths open!

Notwithstanding the diversity of opinions about how the world came into existence and whether the world has purpose and meaning, there is only one source for human understanding—life experience. As Einstein expressed in his *Credo*, religion and science begin from the same place:

> The most beautiful and deepest experience a man can have is the sense of the mysterious. It is the underlying principle of religion as well as all serious endeavors in art and science. He who never had this experience seems to me, if not dead, then at least blind. To sense that behind anything that can be experienced there is a something that our mind cannot grasp and whose beauty and sublimity reaches us only indirectly and as a feeble reflection, this is religiousness. In this sense I am religious. (12)

Some years ago, I heard the following example used to illustrate how scientific and religious attitudes can divide life experience into different domains of human understanding and action. Imagine walking along a beach and coming upon a large and unusual rock. Two different sets of possible questions arise:

- What kind of rock is this? How did it get here? What can be done with it?

- What does it mean that this rock and I are sharing the beach together at this moment in time? What can this moment (or rock) teach me about the meaning of life?

The first set of questions represents science and technology. Knowing the answers enables the control needed to get more or fewer rocks according to one's needs and desires. The second set of questions represents religion and spirituality. They concern the meaning and purpose of the individual and of life.

Religion and spirituality are words frequently used interchangeably. James's religious attitude includes both. Through its communal beliefs and practices, religion describes the nature of the unseen order and offers insights—sometimes commandments—regarding how to harmonize oneself with this order. The individual quest for meaning—the person's spiritual encounter with the world—offers its own insights. Spiritual insights frequently determine just how seriously the person takes what religion has to offer or command. Sometimes, spiritual insights lead a person to try to change the religion into something different.

Archetypes

As distinct ways of understanding and interpreting life experience, science and religion can be situated historically by metaphors such as Athens versus Jerusalem (13) or Plato versus the Prophets (14). According to Rabbi Joseph Soloveitchik, one already can find this juxtaposition within the two biblical creation stories of Adam—two human archetypes—Genesis 1 and 2. "We all know that the Bible offers two accounts of the creation of man," wrote Soloveitchik, "two Adams, two men, two fathers of mankind, two types, two representatives of humanity" (15).

Here are the relevant biblical passages:

Creation of Adam I in Genesis 1:

> Genesis 1.27: And God created man in His image; in the image of God He created him; male and female He created them. 1.28: And God blessed them, and God said to them, "Be fruitful and multiply and fill the earth and subdue it, and rule over the fish of the sea and over the fowl of the sky and over all the beasts that tread upon the earth."

Creation of Adam II in Genesis 2:

> Genesis 2.7: And the Lord God formed man of dust from the ground, and He breathed into his nostrils the breath of life, and man became a living soul. 2.15: Now the Lord God took the man, and He placed him in the Garden of Eden to work it and to guard it. 2.18: And the Lord God said, "It is not good that man is alone; I shall make him a helpmate opposite him." 2.22: And the Lord God built the side that He had taken from man into a woman, and He brought her to man.

Soloveitchik called Adam I the mathematical scientist whose life work is to populate and control the natural world. Adam I (male and female) has a creative nature that reflects having been created in God's *creating* image. In this description, man and woman were dropped off in the world together—establishing community from the beginning—and told to do their thing.

Adam II, on the other hand, has a much more intimate relationship with God—God breathed into his nostrils. The second Adam begins alone in the world—pure subjectivity rather than community. Rather than commanded to control, Adam II's responsibility is to cultivate and take care of the natural world. God is concerned about Adam II's psychological well-being and creates for him a life partner, for which Adam II has to sacrifice part of himself.

"Intellectual curiosity drives them both [Adam I and II] to confront courageously the 'mysterium magnum' of Being," wrote Soloveitchik. But this quest takes them along different paths. The first Adam seeks power and control. The second Adam seeks meaning: "Why is it?" "What is it?" "Who is God?" (15).

Faith for science

Conventionally, the contrast between Athens and Jerusalem or Plato and the Prophets or Adam I and Adam II is described as reason versus faith. That description obscures what seems to me to be a central element in trying to understand the relationship between science and religion, namely, the idea that science, like religion, also requires faith. My scientific friends dislike this claim a lot. They argue that in science, assumptions are necessary; faith is not. Assumptions can be changed; faith cannot. I suggest that some assumptions are so profound and held with such passion that they appear to me to resemble what we typically call faith.

I began to appreciate the idea that science requires faith at the Franklin Institute Science Museum in Philadelphia. The Franklin Institute was originally founded in 1824, and its mission is "to inspire an understanding of and passion for science and technology learning" (16). In the late 1950s, I visited the museum often and became friendly with some of the local college students who watched over the exhibits and put on special shows. One of those students, Richard Rehberg, would let me join him in the science auditorium as he set up the chemistry show. One day as he was setting up, Dick turned to me and said, "Prove you exist."

It took me a while to understand that he was not joking. I devised all sorts of arguments in my favor. Dick responded that my ideas ultimately depended on sense experience. He made me realize that sense experience is not always reliable. Things looked different with and without my glasses. Which was more correct? How could I be sure? What if I were color blind? Maybe I was deceiving myself. At night, in the park, looking at a figure in the distance—is the figure a statue or a real person? This went on for a few weeks. Here I was in a museum dedicated to science and technology—dedicated to knowledge established by observational evidence—and yet unable to provide irrefutable evidence for something of which I was absolutely sure, namely, that I existed. Quite a formative experience for a 14-year-old!

The conversations with Dick Rehberg began my interest in philosophy. Why do you believe what you think you know? Years later, I realized that Dick probably was taking an introductory philosophy course at the time. Most introductory philosophy courses have a section on the seventeenth/eighteenth-century British empiricist philosophers John Locke, George Berkeley, and David Hume. The empiricists questioned what assurance we had beyond our immediate sense experience and memories for "existence or matters of fact." Their realization created a paradox by making the idea of cause and effect—a central tenet of scientific thinking—depend on one's belief that "the course of nature will continue uniformly the same." This belief, wrote Hume, "we take for granted without any proof" (17).

Making the cause-and-effect relationship contingent on the observer's assumption that nature would continue tomorrow the same as today presented a potential challenge to the development of modern science. Science ignored this challenge completely, wrote philosopher Alfred North Whitehead. Instead, we have instinctive faith that "there is an Order of Nature"—instinctive faith that can be attributed to the influence of Western religious belief (18). And Einstein's often-quoted expression, "Science without religion is lame, religion without science is blind," reflects a similar outlook. Einstein attributed human aspiration toward truth and understanding and "faith in the possibility that the regulations valid for the world of existence are rational" to the sphere of religion (19). Science is weak without the religious belief that the world is comprehensible to reason. How ironic!

In summary, science requires faith in the possibility that nature's patterns and structures can be understood. I call this faith in *intelligible* design, in contrast to *intelligent* design. And we share this faith in intelligible design with others who also believe in the uniformity and continuity (repeatability) of the natural world. Indeed, the possibility of sharing—intersubjectivity—permits personal sense experience and memory (what Hume calls the "testimony of the senses") to be transformed from the realm of individual subjectivity into the community's domain of objective knowledge.

In the exchange of subjectivity for intersubjectivity, the individual observer becomes anonymous and the world becomes anyone's—not just mine. From the perspective of the scientific attitude, credible knowledge is repeatable by the individual, continuous with the past, and able to be verified and validated by others: anyone, anywhere, anytime. Given the extent to which humankind has succeeded in populating and controlling the world, science's faith in intelligible design appears to be well justified indeed.

Because credible knowledge depends on sense experience, current beliefs remain open to the possibility of being discovered incorrect, and rejected at some future moment. Science can never achieve unchangeable truth, that is, truth that no further experience will alter. Truth with a capital "T" always remains an ideal point toward which we imagine our knowledge will some day converge. In the meantime, as William James put it,

> We have to live today by what truth we can get today, and be ready tomorrow to call it falsehood. Ptolemaic astronomy, Euclidean space, Aristotelian logic, scholastic metaphysics, were expedient for centuries, but human experience has boiled over those limits, and we now call these things only relatively true, or true within those borders of experience. "Absolutely" they are false. (20)

The truth of science is pragmatic through and through.

Faith for religion

Religion requires a different kind of faith than science but in no way gives up the demand for reason. One of the most provocative images of reason in religion can be found in Rabbi Soloveitchik's description of *halakhic man*. Halakhah refers to Jewish law, which developed over centuries of debate, discussion, and codification regarding how to implement in everyday life the 613 commandments (Talmud Makkot 23b) that can be found in *Torah* (the first five books of the Christian Old Testament).

Soloveitchik analogized the development of halakhah to formulation of a mathematical system—"*a priori* and ideal," whose "necessity flows from its very nature." Like any mathematical system, validity depends on logical rules applied correctly to starting assumptions. The starting assumptions, however, need not correspond to the shared sensory space in which we all live. At the time mathematicians developed non-Euclidian geometry, Soloveitchik reminds us, the world was experienced as fully Euclidean (21). What distinguishes religion from science is not the absence of reason but rather a willingness to accept starting assumptions from outside of shared sensory space.

Finding knowledge outside of shared sensory space sounds like Emily Dickinson's description of a color that human nature feels but science cannot overtake. Poet Walt Whitman offers a similar sentiment in "When I heard the learn'd astronomer":

> When I heard the learn'd astronomer,
> When the proofs, the figures, were ranged in columns
> before me,
> When I was shown the charts and diagrams, to add, divide,
> and measure them,
> When I sitting heard the astronomer where he lectured
> with much applause in the lecture-room,
> How soon unaccountable I became tired and sick,
> Till rising and gliding out I wander'd off by myself,
> In the mystical moist night-air, and from time to time,
> Look'd up in perfect silence at the stars. (22)

Science can extend the boundaries of shared sensory space with instruments such as microscopes and telescopes. Only by turning away from shared sensory space and entering Whitman's perfect silence can the knowledge be acquired that human nature feels and science cannot overtake—discovery by spiritual encounter.

Philosopher Martin Buber described his spiritual encounter as the I–Thou relationship:

> The form that confronts me I cannot experience nor describe; I can only actualize it. And yet I see it, radiant...far more clearly than all the clarity of the experienced world. Not as a thing among the internal things, nor as a figment of the imagination, but as what is present. Tested for its objectivity, the form is not there at all; but what can equal its presence? And it is an actual relation: it acts on me as I act on it.

What then does one experience of the Thou? Nothing at all. For one does not experience it. What, then, does one know of the Thou? Only everything, for one no longer knows particulars. (23)

Where Thou is said, there is no some-thing because Thou has no borders. The person becomes no-one, the world becomes no-thing, and the two fuse together at a moment out of time—the living present. Buber once commented that he believed in the God you can speak to, not in the God that you can speak about (24).

Lack of borders makes possible unified knowledge without contradictions, wrote theologian Rudolf Otto. Rather than multiple, separate, and divided, the experience occurs as an inexpressible All. "This is that, and that is this; here is there and there, here" (25). James describes such experiences as mystical states of consciousness and calls them the root and center of personal religious experience— sources of "illuminations, revelations, full of significance and importance, all inarticulate though they remain" (11). If the scientific attitude corresponds to anyone, anywhere, anytime, then the mystical attitude corresponds to no one, nowhere, out-of-time.

Spirituality/mysticism is one path to James's unseen order of the religious attitude. Religious communities offer a more commonplace source of starting assumptions. These assumptions can be found in every religion, whether the ethical humanist's commitment to the supreme worth of human life or the elaborate revelations of great religious leaders of the past such as Buddha, Krishna, Moses, Jesus, and Mohammed.

Because each religion embraces a different set of revelations and assumptions about the unseen order, each develops these assumptions into a reasoned framework that provides guidance about values, meaning, and purpose of life. Fragmentation is inevitable, and the possibilities are endless. One cannot find THE religious view on any matter; rather, there always are many. The European Network of

Scientific Co-operation on Medicine and Human Rights compiled 120 case studies regarding medical care and biomedical research. In this compilation, they presented perspectives on each case not only from the point of view of international law and ethics (World Medical Association), but also from six different "religious moralities": Catholic, Protestant, Jewish, Muslim, Buddhist, and agnostic (26).

Because of fragmentation, religion requires a credibility process much different from that of science to maintain its unique guidance about values, meaning, and purpose. Credibility in science begins when peer review authorizes that a discovery claim is worthwhile to examine, but the validity of the claim requires testing by the community over time. At the time the discovery claim is made, the idea could be quite different from prevailing beliefs of the community. At the beginning of the process, the outcome will be uncertain. The credibility process in religion, by contrast, requires peer review to certify at the outset that an individual's insights are consistent with the religion's current understanding of itself.

By using the phrase "current understanding of itself," I want to emphasize two points. First, religious beliefs constantly are evolving as the religion's starting assumptions from outside sensory space are tested against contemporary understanding of the world. The evolution of religious thinking can be extraordinary, as exemplified by changes in moral teachings. A short list of areas in which official Christian teaching has changed over the past 2,000 years includes "slavery, usury, religious freedom, human rights, democracy, the right to silence, the role of women, and many aspects in the understanding of marriage and sexuality" (27). Frequently, change begins when the larger religious community questions the credibility of previously accepted ways of thinking.

Second, there are limits to change. "We should examine the Scriptural texts by the intellect," wrote Rabbi Moses Maimonides in *The Guide for the Perplexed*, "after having acquired a knowledge of demonstrative science, and of the true hidden meaning of prophe-cies" (28). Evaluating Scripture by demonstrative science is accept-able, but a religion's understanding of the "true hidden meaning of prophecies" establishes limits. Going beyond those limits means leaving the religion. For instance, such limits were established by the Council of Trent's "Decree concerning the edition, and the use, of the sacred book" as part of the Catholic Church's response to the Protestant Reformation:

> [I]n order to restrain petulant spirits, It [the Holy Synod] decrees, that no one, relying on his own skill, shall,—in matters of faith, and of morals pertaining to the edification of Christian doctrine,—wresting the sacred Scripture to his own senses, presume to interpret the said sacred Scripture contrary to that sense which holy mother Church...(29)

In contemporary times, preventing a book from being published and distributed is difficult to accomplish. It still is possible, however, to tell members of your religious group, "This reading material is deadly for your soul. Keep away!" (30).

Unlike the scientific attitude that never reaches absolute Truth (that which no further experience will change), the religious attitude unconditionally accepts certain Truths from the past. These prior Truths correspond to Soloveitchik's starting assumptions from out-side sensory space and to the unseen order in James's definition of the religious attitude. By accepting these Truths as they are understood at any given moment, individuals choose to be part of a particular religious community.

Intelligent design

Intelligent design (ID) offers a good example with which to distin-guish faith in science from faith in religion. The ID movement has received widespread attention because of the legal battles over what should be taught in the science curriculum regarding evolution. The question has been turned into a political issue. In the state where I work, the Republican platform includes the following section about "Theories of Origin":

> We support the objective teaching and equal treatment of scientific strengths and weaknesses of scientific theories, including Intelligent Design. We believe theories of life origins and environmental theories should be taught as scientific theory not scientific law; that social studies and other curriculum should not be based on any one theory. (31)

The pro-Creation/pro-ID/anti-evolution educational lobby tries to dilute education about evolution whenever their candidates win control of local and state school boards that determine science educational standards. The National Center for Science Education (NCSE), on the other hand, works against any decrease in teaching about evolution. The goal of the NCSE is to educate the "press and public about the scientific, educational, and legal aspects of the creation and evolution controversy" and to defend "good science education at local, state, and national levels" (32). That the anti-evolution lobby in the United States has succeeded to some extent may explain why national surveys of U.S. adults (most recently in 2005) show that only about 40% accept evolution. Approximately another 40% reject evolution, and the rest are unsure. Public acceptance of evolution in the United States is lower than in most other countries (33).

Superficially, ID sounds scientific, arguing that "neo-Darwinism has failed to explain the origin of the highly complex information systems and structures of living organisms, from the first cells to new body plans" (34). Underlying the ID argument is a discovery claim called *irreducible complexity*. Initially based on a statistical analysis of protein structure and function (35), irreducible complexity denies the possibility of common ancestry of life forms as described by modern evolutionary biology. ID proponents say that in light of the limits imposed by irreducible complexity, the possibility of evolution depends on intervention of a hypothetical force outside the known laws of nature.

The problem of explaining highly complex information systems and structures of living organisms is not new to biology. Biologists in the seventeenth century could not imagine organs and tissues arising from a single cell—the fertilized ovum. The hypothesis called *preformationism* resolved this dilemma. According to preformationism, development simply reflected an unfolding and increase in size of an organism that already preexisted within germ cells. Perhaps the most famous example of this idea is the 1694 drawing by one of the inventors of the modern microscope, Nicolas Hartsoeker, showing a miniature human within a sperm, that is, a *homunculus* (36). Preformationism was rejected a century later, when physiologist Friedrich Wolff demonstrated that development of a single cell could indeed give rise to the entire set of new tissues and organs. Without an understanding of the underlying processes, Wolff invoked a theoretical, organizational force—the *vis essentialis*—to account for his observations (37).

The *vis essentialis* of eighteenth-century developmental biology resembles superficially the idea of intelligent design. Wolff was a contemporary of theologian William Paley, whose 1802 book *Natural Theology* was forerunner to the ID movement. Nevertheless, the *vis essentialis* was understood by developmental biologists as a "place holder" awaiting further elucidation and was not to be taken literally. Research continued. Another 200 years of cell biology, biochemistry, and genetics has provided the contemporary account of development that effectively replaces the *vis essentialis*.

Supporters of ID are not interested in further elucidation of irreducible complexity. Instead, they appear to be satisfied that they have arrived at the Truth of the matter. Their faith comes from outside of shared sensory space. Rather than a placeholder awaiting further elucidation, they take the idea of ID literally. When ID supporters experience "the sublimity and marvelous order which reveal themselves both in nature and in the world of thought," they do not find Einstein's "cosmic religious feeling [but] no anthropomorphic conception of God" (38). Rather, they follow the glance of Isaiah 40.26, "Lift up your eyes on high and see—Who created these." Whatever one might think of the merit or failure of ID in terms of religion, its underlying assumptions situate the idea of ID outside of science. Consequently, ID has no place in science education.

Michael Ruse suggests in *The Evolution-Creation Struggle* that the struggle is "not a science-versus-religion conflict but a religion-versus-religion conflict" (39)—faith in science going head to head with faith in religion. Understood as a faith-versus-faith struggle, competing messianic world views (Ruse says *millenialist*) are at stake. On the one hand, we have human redemption by God—the belief that the world has intrinsic purpose and meaning. On the other hand, we have redemption of the world by humans—the belief that whatever purpose and meaning the world develops, we humans will create (*global existentialism*).

Given the technological power of modern science, we are no longer mere spectators in the evolutionary process. We have entered a kind of postevolutionary phase. Instead of adapting to environmental change, we are changing the world's environment. Instead of random mutation and selection, modern techniques of molecular, cellular, and developmental biology have made possible manipulation of organisms, including ourselves, in a highly specific fashion at the genomic level. The stakes are very high indeed.

Complementarity

In their 1999 report on evolution, creationism, and ID, a committee from the National Academies wrote:

> Scientists, like many others, are touched with awe at the order and complexity of nature. Indeed, many scientists are deeply religious. But science and religion occupy two separate realms of human experience. Demanding that they be combined detracts from the glory of each. (40)

Separate and separated are not the same thing. There is only one world, although the world can be experienced in different ways, as shown by the example given earlier of encountering a rock on the beach. Science and religion may be separate, but they cannot remain separated. Some kind of interaction is inevitable.

Bicycle riding frequently is offered as a metaphor to describe the interaction of science and religion. Having a bike makes riding possible. Other factors influence the direction in which the rider will choose to go. Science provides the technology for doing things. Religion provides the values to determine what things should be done. Einstein conveys this relationship in his description:

> Objective knowledge provides us with powerful instruments for the achievements of certain ends, but the ultimate goal itself and the longing to reach it must come from another source....Intelligence makes clear to us the interrelation of means and ends. But mere thinking cannot give us a sense of the ultimate and fundamental ends. To make clear these fundamental ends and valuations, and to set them fast in the emotional life of the individual, seems to me precisely the most important function which religion has to form in the social life of man. (19)

Rather than "Science without religion is lame, religion without science is blind," Einstein might just have well have said, "Science without religion is blind, religion without science is lame."

The bicycle metaphor emphasizes the idea that scientific and religious attitudes can make up for each other's deficiencies— complementary in a functional sense (41). I will now suggest that these two attitudes are complementary in a more profound sense of what Niels Bohr called *complementarity*. Bohr introduced complementarity in 1927 at the International Congress of Physics held in Como, Italy, to "harmonize the different views, apparently so divergent," concerning the wave and particle explanations of light (42).

Bohr's use of the word complementarity in quantum physics has a much different connotation than the everyday experience of complementary descriptions. In everyday life, we all have the experience that single views of three-dimensional objects are incomplete. Clothing stores typically have multiple mirrors so that potential buyers can see how they look from multiple perspectives—deciding "how I look" requires more than just a head-on view. To appreciate fully the rock on the beach, one has to visit it under different conditions of light and weather and to look at it from all sides. Likewise, a single map of geographic space will likely be incomplete. Different types of geographic maps—for example, resolution, topography, and points of interest—describe the landscape from different perspectives (43). We take it for granted that different perspectives or maps taken together converge into a more complete image and understanding of the object under examination. The perspectives change; the identity of the object remains fixed.

Closer to but still different from Bohr's notion of complementarity are observations of hierarchical systems. Trying to understand the rock on the beach may require not only studying its surface features, but also disrupting the surface to examine the contents within. The objects of scientific investigation often are hierarchical systems that can be viewed either as organized wholes or in terms of individual parts. Holistic studies focus on the system's organization and features that emerge from that organization. Reductionist studies focus on the individual parts. If carrying out reductionist studies requires destroying the intact object, then studying both the whole and the parts at the same time becomes impossible. Nevertheless, the identity of the object is assumed constant when one makes observations on both sides of a hierarchical boundary.

A third example of complementary descriptions comes from William James's metaphor of a sculptor working on a block of stone. James uses this metaphor to describe the constitution of everyday life experience in the stream of consciousness:

> The mind is at every stage a theater of simultaneous possibilities. Consciousness consists in the comparison of these with each other, the selection of some, and the suppression of the rest by the reinforcing and inhibiting agency of attention.... The mind, in short, works on the data it receives very much as a sculptor works on his block of stone. In a sense the statue stood there from eternity. But there were a thousand different ones beside it and the sculptor alone is to thank for having extricated this one from the rest. (44)

The artistic circumstances in which each of these thousand different statues arises are mutually exclusive. Sculptor and statue make up an interacting unit. The block of stone creates limits, but the object has no fixed identity apart from the work of the sculptor.

In an interview conducted shortly before his death, Bohr implied that some of his ideas about complementarity may have been derived at least in part from James's explanation of consciousness (45). The problem Bohr confronted in 1927 concerned the failure of classical physics to explain the discoveries of atomic behavior. The nature of light presented an experimental and theoretical puzzle because two sets of evidence and two theories had become associated with light propagation. One theory understood light as light waves; the other, as light quanta (particles) (46). Wave and particle observations of light were mutually exclusive experimentally and theoretically, and yet they accounted for equally important features of the light phenomenon.

Bohr explained the duality of light by changing the conceptual framework for understanding the phenomenon. He argued that at the quantum level, there could be no distinction between the object and the experimental circumstances that permitted the object to be observed. Unlike the conventional notions of complementary perspectives in which observer and object remain separate, in complementarity, observer and object make up an interacting unit like the sculptor and the block of stone (47). Views obtained under different circumstances are required for a comprehensive understanding of the phenomenon under investigation. However, these views are mutually exclusive in the sense that they cannot be brought into direct contradiction with each other and cannot disprove each other. Rather, they exist side by side, each adequate within its own experimental, observational framework. When he was knighted, Bohr symbolically expressed his commitment to complementarity by choosing the *yin-yang* symbol as his family crest.

Bohr did not believe that complementarity was limited to quantum physics. He argued that the ability of complementarity to solve an unexpected paradox of physics should encourage an attempt to use this approach in other domains of experience, including science and religion:

A whole new background for the relationship between scientific research and religious attitude has been created by the modern development of physics....Materialism and spiritualism, which are only defined by concepts taken from each other, are two aspects of the same thing. (quoted in 47)

Figure 6.1 diagrams the complementarity of science and religion as "two aspects of the same thing" by combining these two attitudes into Bohr's yin-yang family crest. Life for the person begins in the me, here, now of everyday life experience. The filters (interpretive/experiential frameworks) of the religious and scientific attitudes reveal distinct domains of knowledge that cannot be seen or inferred or negated from the other perspective. The religious attitude gives us James's unseen order to which the individual seeks to conform. The scientific attitude gives us the anyone, anywhere, anytime of inter-subjectivity. These domains are separate, but they cannot remain sep-arated. Rather, they merge into a holistic yin-yang framework that cannot be harmonized or resolved further. They exist in dynamic tension—constantly bouncing off each other and inevitably offering

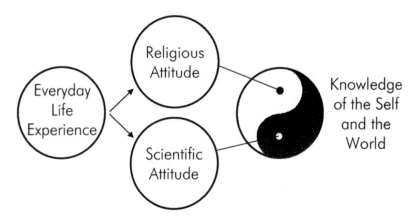

Figure 6.1. Complementarity of the Religious and Scientific Attitudes

different types of answers to the fundamental questions about the self and the world.

Dialogue

Bohr suggested that another aspect of life experience, "the foundation of so-called ethical values," also could be understood in complementary terms of actor and spectator:

> We shall in the first place ask about the scope of such concepts as justice and charity, the closest possible connection of which is attempted in all human societies. Still it is evident that a situation permitting unambiguous use of accepted judicial rules leaves no room for the free display of charity. [It] is equally clear that compassion can bring everyone into conflict with any concisely formulated idea of justice. We are here confronted with complementary relationships inherent in the human position, reminding us that in the great drama of existence we are ourselves both actors and spectators. (48)

When complementarity first was put forth as a possible framework for relating religion and science, actor and spectator were described as representative of scientific and religious perspectives, respectively (49). The problem with this characterization is its asymmetry. In the me, here, now of everyday experience, we all are actors and spectators. The religious attitude is fully prepared to incorporate the spectator perspective. "True science," Pope Pius XII told the Pontifical Academy of Sciences, "discovers God in an ever-increasing degree—as though God were waiting behind every door" (50). The scientific attitude, by contrast, seeks to eliminate from discovery claims just those features that are specific to the actor perspective.

The lack of symmetry between science and religion with respect to spectator and actor perspectives presents a significant problem for what has become known as the dialogue between science and religion. I came to appreciate this problem firsthand in 1992 at a symposium called "Darwinism: Science or Philosophy." The discovery claim of irreducible complexity made its first appearance at that symposium. In a lecture titled "Radical Intersubjectivity: Why Naturalism Is an Assumption Necessary for Doing Science," I concluded that "if it can't be measured or counted or photographed, then it can't be science—even if it's important" (51). In the discussion period that followed, Philip Johnson—one of the leaders of the ID movement—suggested that my comments indicated I was assigning second-class status to religious intuitions. Almost 10 years later, I attended a workshop at Oxford University on the topic of "Knowledge, Reality and Imagination: Critical Realism and Its Alternatives in Science and Religion." I sensed frustration on the part of many participants similar to that expressed by Johnson regarding what they saw as an openness of religious thinking to scientific discoveries that was not reciprocated by science. This lack of reciprocity has the appearance of being formalized in the goals of the American Association for the Advancement of Science program called "Dialogue on Science, Ethics, and Religion: Promoting a Public Conversation":

> To facilitate communication between scientific and religious communities...contribute to the level of scientific understanding in religious communities...[and] promote multidisciplinary education and scholarship of the ethical and religious implications of advancements in science and technology. (52)

Currently, the dialogue between science and religion tends to be one-sided—science informing religion. To go beyond this level will require accepting scientific and religious attitudes as complementary in the holistic sense of dynamic tension shown in figure 6.1.

Humility

In the Talmud, two famous schools of Jewish religious thought usually are presented as being in conflict with each other: the school of Rabbi Hillel versus the school of Rabbi Shammai. In one case, the point of dispute concerned how a person should interact with a bride on her wedding day (*Ketubot* 16b–17a). The school of Shammai offered the scientific view—how she appears to anyone. The school of Hillel offered a more spiritual view—how she appears to her husband. In another example, the two schools debated a point of law (the text does not really tell us what point was under consideration) for three years until a heavenly voice announced:

> "Both are the words of the living God, but the halakhah is in agreement with the rulings of Beth Hillel." Since, however, "both are the words of the living God" what was it that entitled Beth Hillel to have the halakhah fixed in agreement with their rulings? Because they [Beth Hillel] were kindly and modest, they studied their own rulings and those of Beth Shammai, and were even so [humble] as to mention the actions of Beth Shammai before [their own]. (*Talmud Erubin* 13b) (53)

Complementary views—both correct, both the words of God. A choice has to be made. In this case, the text offers a clue as to why the halakhah went with the school of Hillel—not because they had the right answer but because they were more modest.

"What I see in nature," wrote Einstein, "is a magnificent structure that we can comprehend only very imperfectly, and that must fill a thinking person with a feeling of humility" (quoted in 54).

Recognizing the limitations of our understanding is one of the most important insights from Bohr's complementarity. Perhaps there are no single correct answers. Awareness of this limitation would be helpful when considering the messianic implications of the conflict between science and religion. Personally, I am not expecting redemption by God anytime soon, but I am not so confident about redemption by humans, either. Given the global level of poverty, disease, and depletion of environmental resources, I worry that the scientific attitude conflates power with progress. Perhaps solving global problems will require the scientific *and* religious attitudes—both types of faith—rather than one or the other.

AFTERWORD

My seventh-grade science teacher, Mr. Perkins, had us keep notebooks that were part journal and part a record of our experiments. I still have my notebook. At the top of the first page, dated Tuesday, September 11, 1956, I wrote, "Science is serious play." Mr. Perkins used that expression a lot, and I believed him. Many years later, I still think he was correct. In some respects, writing this book has been my attempt to share with others a better appreciation for the idea that "science is serious play."

Before I decided on the title "Everyday Practice of Science," I considered three other possibilities. The first was "Intentionality of Science," to emphasize that scientists do not passively observe the world as it is, but rather actively impose meaning on everyday life experience through their thought styles. The second was "Shaping Scientific Thought," to emphasize that scientific discoveries do not just happen, but rather emerge dynamically. The third was "Circles of Science," to emphasize that practice of science involves two sets of conversations—one between the researcher and the world to be studied and the other between the researcher and other members of the research community—the circles of discovery and credibility.

Implicit in all three titles was an opportunity to correct what I see as key distortions in the understanding of everyday practice that are associated with the linear model of science:

- Researchers observe and collect facts about the world.

- Researchers use the scientific method to make discoveries.

- Researchers are dispassionate and objective observers.

From chapters 1–3, the reader should appreciate better that everyday practice of science depends on intuition and passion as much as logic and objectivity; that instead of the anonymity of textbook science, real-life researchers experience excitement and adventure; and that scientists are engaged in a highly communal activity, not working alone and isolated. I tried to make clear the role of noticing the unexpected in discovery; the challenge at the frontiers of knowledge to distinguish data from noise; the ambiguity of the scientific method, except in research publications, where it becomes the plot; the process by which discovery claims become credible; and the importance of understanding how thought styles influence practice. Success in science requires commitment and entails risks. Objectivity is a function of the community, not the individual. Because of the importance of the community, diversity of research interests and assumptions—including differences in gender and in racial and economic background of investigators—enhances scientific exploration of the world.

My description of "Everyday Practice of Science" goes beyond intentionality, shaping, and circles. In chapters 4 and 5, I offered readers insights into the social and political dimensions of practice, which involve society as a whole—science and society. The practice of science requires making value judgments about what to do, when to do it, how to do it, how to pay for it, and so forth. For this reason, the dual review criteria used by the National Science Foundation to evaluate its research applications should be part of every grant proposal that requests public funding. The first criterion concerns intellectual merit and can be thought of as dependent on scientific opportunity. The second concerns the broader category of societal needs, which includes education, diversity, and infrastructure. Balancing opportunities versus needs cannot be accomplished successfully without understanding the ambiguities as well as the possibilities inherent in practice. Similarly, effective science policy decisions to regulate responsible conduct, conflict of interest, and the ethics of human research require a realistic appreciation of what practice entails. Society cannot have the benefits of scientific research without the risks, both anticipated and unanticipated.

Finally, my goal in chapter 6 was to contrast the practices of science and religion and explain their relationship as a reflection of two human archetypes—two types of faith. If chapter 6 accomplishes its purpose, then the reader will understand the idea of complementarity as a holistic framework within which science and religion interact dynamically. Neither of these ways of knowing the world is about to relinquish its practices or defer to the other.

References

Chapter 1: Practicing Science

1. Sagan, C. (1993) Why we need to understand science. *Mercury.* 22, 52–55.
2. Medawar, P. (1982) *Pluto's Republic.* (Oxford University Press, New York, N.Y.).
3. Koshland, D. E., Jr. (1996) How to get paid for having fun. *Annu Rev Biochem.* 65, 1–13.
4. Merton, R. K. (1942) The normative structure of science. Reprinted in *The Sociology of Science: Theoretical and Empirical Investigations.* Ed. Merton, R. K., and Storer, N. W. (University of Chicago Press, Chicago, Ill., 1973), pp. 267–278.
5. Brenner, S., Jacob, F., and Meselson, M. (1961) An unstable intermediate carrying information from genes to ribosomes for protein synthesis. *Nature.* 190, 576–581.
6. Jacob, F. (1988) *The Statue Within.* (Basic Books, New York, N.Y.).
7. deSolla Price, D. J. (1981) The development and structure of the biomedical literature. In *Coping with the Biomedical Literature.* Ed. Warren, K. S. (Praeger, New York, N.Y.), pp. 3–16.
8. Pickering, A. (1992) *Science as Practice and Culture.* (University of Chicago Press, Chicago, Ill.).
9. Fuller, S. (2002) *Social Epistemology.* (Indiana University Press, Bloomington, Ind.).
10. Kuhn, T. S. (1962) *The Structure of Scientific Revolutions.* (University of Chicago Press, Chicago, Ill.).
11. Kuhn, T. S. (1979) Objectivity, value judgment, and theory choice. In *The Essential Tension: Selected Studies in Scientific Tradition and Change.* (University of Chicago Press, Chicago, Ill.), pp. 320–339.
12. Piaget, J. (1970) *Genetic Epistemology.* (W. W. Norton, New York, N.Y.).

13. Fleck, L. (1935) *Genesis and Development of a Scientific Fact.* (Reprint: University of Chicago Press, Chicago, Ill., 1979).

14. Polanyi, M. (1958) *Personal Knowledge: Towards a Post-critical Philosophy.* (University of Chicago Press, Chicago, Ill.).

15. Holton, G. (1973) *Thematic Origins of Scientific Thought: Kepler to Einstein.* (Harvard University Press, Cambridge, Mass.).

16. Martin, S. (1996) *Picasso at the Lapin Agile and Other Plays.* (Grove Press, New York, N.Y.).

17. Rosser, S. V. (2001) Are there feminist methodologies appropriate for the natural sciences and do they make a difference? In *The Gender and Science Reader.* Ed. Lederman, M., and Bartsch, I. (Routledge, New York, N.Y.), pp. 123–144.

18. Gould, S. J. (1981) *The Mismeasure of Man.* (W. W. Norton, New York, N.Y.).

19. Bayer, R. (1981) *Homosexuality and American Psychiatry.* (Basic Books, New York, N.Y.).

20. Hayes, W. (1982) Max Ludwig Henning Delbrück. *Biographical Memoirs of the Fellows of the Royal Society.* 28, 58–90.

21. Dulbecco, R. (1999) Phone interview with the author, 14 January.

22. Schutz, A. (1967) *The Phenomenology of the Social World.* (Northwestern University Press, Evanston, Ill.).

23. Goldstein, J. L. (2007) Creation and revelation: Two different routes to advancement in the biomedical sciences. *Nat Med.* 13, 1151–1154.

24. Ziman, J. M. (1968) *Public Knowledge: An Essay Concerning the Social Dimension of Science.* (Cambridge University Press, Cambridge, U.K.).

25. James, W. (1907) Pragmatism's conception of truth. Reprinted in *Pragmatism and the Meaning of Truth.* (Harvard University Press, Cambridge, Mass., 1975), pp. 95–113.

26. Baier, A. C. (1997) *The Commons of the Mind.* (Open Court, Chicago, Ill.).

27. Frankel, M. S. (1993) *Viewing Science and Technology through a Multicultural Prism.* (American Association for the Advancement of Science, New York, N.Y.).

28. Mervis, J. (2006) U.S. education policy. NSF board wades into swirling debate on school reform. *Science.* 312, 45.

29. Lederman, L. (2000) A plan, a strategy for K-12. In *Who Will Do the Science of the Future? A Symposium on Careers of Women in Science.*

Ed. National Academy of Sciences (National Academy Press, Washington, D.C.), pp. 7–11.

30. Conant, J. B. (1951) *Science and Common Sense.* (Yale University Press, New Haven, Conn.).

31. Aldridge, B. G. (1992) Project on scope, sequence, and coordination: A new synthesis for improving science education. *J Sci Educ Technol.* 1, 13–21.

Chapter 2: Discovery

1. Freeman, R., Weinstein, E., Marincola, E., Rosenbaum, J., and Solomon, F. (2001) Competition and careers in biosciences. *Science.* 294, 2293–2294.

2. Nobel Foundation (1974) *The Nobel Prize in Physiology or Medicine 1974.* http://nobelprize.org/nobel_prizes/medicine/laureates/1974/.

3. Klug, J. (1998) From hypertension to angina to Viagra. *Mod Drug Discov.* 1, 31–38.

4. Comroe, J. H., Jr., and Dripps, R. D. (1974) Ben Franklin and open heart surgery. *Circ Res.* 35, 661–669.

5. Comroe, J. H., Jr., and Dripps, R. D. (1976) Scientific basis for the support of biomedical science. *Science.* 192, 105–111.

6. Ziman, J. M. (1968) *Public Knowledge: An Essay Concerning the Social Dimension of Science.* (Cambridge University Press, Cambridge, U.K.).

7. Leinster, M. (1960) *The Wailing Asteroid.* (Avon Books, New York, N.Y.).

8. Oxford English Dictionary Online (1989) http://dictionary.oed.com/.

9. Roberts, R. M. (1989) *Serendipity: Accidental Discoveries in Science.* (John Wiley & Sons, New York, N.Y.).

10. Schilling, H. K. (1958) A human enterprise. *Science.* 127, 1324–1327.

11. Holton, G. (1973) *Thematic Origins of Scientific Thought: Kepler to Einstein.* (Harvard University Press, Cambridge, Mass.).

12. Koshland, D. E. J. (1990) The addictive personality. *Science.* 250, 1193.

13. Plato (380 B.C.E.) *Meno.* http://classics.mit.edu/Plato/meno.html.

14. Doyle, A. C. S. (1905) Silver blaze. In *The Complete Sherlock Holmes.* (Doubleday, Garden City, N.Y.), pp. 335–350.

15. Wilson, A. T., and Calvin, M. (1955) The photosynthetic cycle: CO_2 dependent transitions. *J Am Chem Soc.* 77, 5948–5957.

16. Polanyi, M. (1966) *The Tacit Dimension*. (Reprint: Peter Smith Publishers, Gloucester, Mass., 1983).

17. Hayes, W. (1982) Max Ludwig Henning Delbrück. *Biographical Memoirs of the Fellows of the Royal Society*. 28, 58–90.

18. Dulbecco, R. (1999) Phone interview with the author, 14 January.

19. Peirce, C. P. (1903) Harvard Lectures on pragmatism. Reprinted in *Collected Papers of Charles Sanders Peirce: Vol. 5*. Ed. Hartshorne, C., Weiss, P., and Burks, A. (Harvard University Press, Cambridge, Mass., 1935), pp. 188–189.

20. Roseman, S. (1970) The synthesis of complex carbohydrates by multiglycosyltransferase systems and their potential function in intercellular adhesion. *Chem Phys Lipids*. 5, 270–297.

21. Roseman, S. (1970) Letter to the author, November 6.

22. Bernard, C. (1865) *An Introduction to the Study of Experimental Medicine*. (Reprint: Dover, New York, N.Y., 1957).

23. Reader, J. (1981) *Missing Links: The Hunt for Earliest Man*. (HarperCollins, New York, N.Y.).

24. Couzin, J. (2003) Estrogen research: The great estrogen conundrum. *Science*. 302, 1136–1138.

25. Turgeon, J. L., McDonnell, D. P., Martin, K. A., and Wise, P. M. (2004) Hormone therapy: Physiological complexity belies therapeutic simplicity. *Science*. 304, 1269–1273.

26. Manson, J. E., Allison, M. A., Rossouw, J. E., Carr, J. J., Langer, R. D., et al. (2007) Estrogen therapy and coronary-artery calcification. *N Engl J Med*. 356, 2591–2602.

27. Popper, K. R. (1959) *The Logic of Scientific Discovery*. (Basic Books, New York, N.Y.).

28. Kornberg, A. (1989) *For the Love of Enzymes: The Odyssey of a Biochemist*. (Harvard University Press, Cambridge, Mass.).

29. Grinnell, F. (1975) Cell attachment to a substratum and cell surface proteases. *Arch Biochem Biophys*. 169, 474–482.

30. Grinnell, F. (1976) Cell spreading factor. Occurrence and specificity of action. *Exp Cell Res*. 102, 51–62.

31. Fleck, L. (1935) *Genesis and Development of a Scientific Fact*. (Reprint: University of Chicago Press, Chicago, Ill., 1979).

32. Kuhn, T. S. (1962) *The Structure of Scientific Revolutions*. (University of Chicago Press, Chicago, Ill.).

33. Engelhardt, H. T. J. (1986) *The Foundations of Bioethics*. (Oxford University Press, New York, N.Y.).

34. Kornberg, A. (1991) Understanding life as chemistry. *Clin Chem*. 37, 1895–1899.

35. McClintock, B. (1983) Nobel Banquet Speech—December 10, 1983. http://nobelprize.org/nobel_prizes/medicine/laureates/1983/mcclintock-speech.html.

36. Jacob, F. (1988) *The Statue Within*. (Basic Books, New York, N.Y.).

37. Levi-Montalcini, R. (1988) *In Praise of Imperfection*. (Basic Books, New York, N.Y.).

38. Gurwitsch, A. (1964) *The Field of Consciousness*. (Duquesne University Press, Pittsburgh, Penn.).

39. Heelan, P. A. (1989) After experiment: Realism and research. *Am Philos Q*. 26, 297–308.

40. Piaget, J. (1970) *Genetic Epistemology*. (W. W. Norton, New York, N.Y.).

41. Gardner, H. (1993) *Creating Minds*. (Basic Books, New York, N.Y.).

42. Stephan, P. E., and Levin, S. G. (1993) Age and the Nobel Prize revisited. *Scientometrics*. 28, 387–399.

43. de Solla Price, D. (1983) The science/technology relationship, the craft of experimental science, and policy for the improvement of high technology innovation. In *Proceedings of the Workshop on the Role of Basic Research in Science and Technology: Case Studies in Energy R&D*. (National Science Foundation, Washington, D.C.), pp. 225–254.

Chapter 3: Credibility

1. Gurwitsch, A. (1964) *The Field of Consciousness*. (Duquesne University Press, Pittsburgh, Penn.).

2. Goodenough, U. (2000) *The Sacred Depths of Nature*. (Oxford University Press, New York, N.Y.).

3. The White House Office of the Press Secretary (2000) *Remarks... on the Completion of the First Survey of the Entire Human Genome Project*. http://www.genome.gov/10001356.

4. Einstein, A. (1954) The world as I see it. In *Ideas and Opinions*. (Crown Publishers Inc., New York, N.Y.), pp. 8–11.

5. Miró, J. (1918) Letter to J. F. Ràfols, 1 August. (Exhibition, Joan Miró 1917–1934, *The Birth of the World*, Centre Pompidou, Paris, France).

6. Natanson, M. (1970) *The Journeying Self: A Study in Philosophy and Social Role*. (Addison-Wesley, Reading, Mass.).

7. Schutz, A. (1970) *On Phenomenology and Social Relations*. (University of Chicago Press, Chicago, Ill.).

8. Ruse, M. (1986) *Taking Darwin Seriously: A Naturalistic Approach to Philosophy*. (Basil Blackwell, Oxford, U.K.).

9. Fleck, L. (1935) *Genesis and Development of a Scientific Fact*. (Reprint: University of Chicago Press, Chicago, Ill., 1979).

10. Gardner, H. (1993) *Creating Minds*. (Basic Books, New York, N.Y.).

11. Ziman, J. M. (1968) *Public Knowledge: An Essay Concerning the Social Dimension of Science*. (Cambridge University Press, Cambridge, U.K.).

12. Select Committee on Science and Technology (2004) *The Origin of the Scientific Journal and the Process of Peer Review*. http://www.publications.parliament.uk/pa/cm200304/cmselect/cmsctech/399/399we23.htm.

13. James, W. (1907) Pragmatism's conception of truth. Reprinted in *Pragmatism and the Meaning of Truth*. (Harvard University Press, Cambridge, Mass., 1975), pp. 95–113.

14. Nass, S. J., & Stillman, B. W., eds. (2003) *Large-Scale Biomedical Science: Exploring Strategies for Future Research*. (National Academies Press, Washington, D.C.).

15. Venter, J. C., Adams, M. D., Myers, E. W., Li, P. W., Mural, R. J., et al. (2001) The sequence of the human genome. *Science*. 291, 1304–1351.

16. Lander, E. S., Linton, L. M., Birren, B., Nusbaum, C., Zody, M. C., et al. (2001) Initial sequencing and analysis of the human genome. *Nature*. 409, 860–921.

17. SSC Scientific and Technical Electronic Repository (1995) *The Superconducting Super Collider Project*. http://www.hep.net/ssc/.

18. Medawar, P. (1988) *Memoir of a Thinking Radish*. (Oxford University Press, Oxford, U.K.).

19. Dyson, F. (2004) A meeting with Enrico Fermi. *Nature*. 427, 297.

20. Knorr-Cetina, K. D. (1981) *The Manufacture of Knowledge*. (Pergamon Press, Oxford, U.K.).

21. Latour, B., and Woolgar, S. (1979) *Laboratory Life: The Construction of Scientific Facts.* (Princeton University Press, Princeton, N.J.).
22. Jacob, F. (1988) *The Statue Within.* (Basic Books Inc., New York, N.Y.).
23. Brenner, S., Jacob, F., and Meselson, M. (1961) An unstable intermediate carrying information from genes to ribosomes for protein synthesis. *Nature.* 190, 576–581.
24. Popper, K. R. (1959) *The Logic of Scientific Discovery.* (Basic Books Inc., New York, N.Y.).
25. Hanson, N. R. (1961) Is there a logic of scientific discovery? In *Current Issues in Philosophy of Science.* Ed. Feigl, H., and Maxwell, G. (Holt, Rinehart and Winston, Inc., New York, N.Y.). pp. 20–35.
26. Marton, L. (1941) The electron microscope: A new tool for bacteriological research. *J Bacteriol.* 41, 397–413.
27. Meyer, S. C. (2004) The origin of biological information and the higher taxonomic categories. *Proc Biol Soc Wash.* 117, 213–239.
28. Council of the Biological Society of Washington (2002) *Statement.* http://www.biolsocwash.org/id_statement.html.
29. *Daubert v. Merrell Dow Pharmaceuticals* (92–102), 509 U.S. 579 (1993).
30. Yalow, R. S. (1992) The Nobel Lecture—Radioimmunoassay: A probe for fine structure of biologic systems. *Scand J Immunol.* 35, 4–23.
31. Science.ca (2001) *Profile: Michael Smith.* http://www.science.ca/scientists/scientistprofile.php?pID=18.
32. Kohler, G., and Milstein, C. (1997) The Landmarks Interviews: Specifically, monoclonal antibodies. *J NIH Res.* 9, 38–44.
33. Hull, D. L. (1988) *Science as a Process.* (University of Chicago Press, Chicago, Ill.).
34. National Institutes of Health (2004) *NIH Announces Updated Criteria for Evaluating Research Grant Applications.* http://grants.nih.gov/grants/guide/notice-files/NOT-OD-05-002.html.
35. National Science Foundation (2006) Academic research and development. In *Science and Engineering Indicators 2006.* http://www.nsf.gov/statistics/seind06/c5/c5i.htm.
36. Grinnell, F., Billingham, R. E., and Burgess, L. (1981) Distribution of fibronectin during wound healing in vivo. *J Invest Dermatol.* 76, 181–189.
37. Wysocki, A. B., Staiano-Coico, L., and Grinnell, F. (1993) Wound fluid from chronic leg ulcers contains elevated levels of

metalloproteinases MMP-2 and MMP-9. *J Invest Dermatol.* 101, 64–68.

38. Grinnell, F., Zhu, M., and Parks, W. C. (1998) Collagenase-1 complexes with alpha2-macroglobulin in the acute and chronic wound environments. *J Invest Dermatol.* 110, 771–776.

39. Welch, G. R. (1999) Paul A. Srere (1925–1999). *Trends Biochem Sci.* 24, 377–378.

40. Srere, P. (1970) I-Ching and the citric acid cycle. Unpublished manuscript/seminar notes.

41. Nobel Foundation (1953) *Presentation Speech: The Nobel Prize in Physiology or Medicine 1953.* http://www.nobel.se/medicine/laureates/1953/press.html.

42. Nobel Foundation (1997) *Presentation Speech: The Nobel Prize in Physiology or Medicine 1997.* http://www.nobel.se/medicine/laureates/1997/presentation-speech.html.

43. Klotz, I. M. (1990) How to become famous by being wrong in science. *Int J Quant Chem.* 24, 881–890.

44. Donahoe, F. J. (1969) "Anomalous" water. *Nature.* 224, 198.

45. Lippincott, E. R., Stromberg, R. R., Grant, W. H., and Cessac, G. L. (1969) Polywater. *Science.* 164, 1482–1487.

46. Hildebrand, J. H. (1970) "Polywater" is hard to swallow. *Science.* 168, 1397.

47. Klotz, I. M. (1985) *Diamond Dealers and Feature Merchants.* (Birkhauser, Boston, Mass.).

48. Derjaguin, B. V., and Churaev, N. V. (1973) Nature of "anomalous water." *Nature.* 244, 430–431.

Chapter 4: Integrity

1. Committee on Assessing Integrity in Research Environments, National Research Council, Institute of Medicine (2002) *Integrity in Scientific Research: Creating an Environment That Promotes Responsible Conduct.* (National Academies Press, Washington, D.C.).

2. President George W. Bush (2001) *President Discusses Stem Cell Research.* http://www.whitehouse.gov/news/releases/2001/08/20010809-2.html.

3. World Medical Association (2004) *Ethical Principles for Medical Research Involving Human Subjects.* http://www.wma.net/e/policy/b3.htm.

4. Committee on Guidelines for Human Embryonic Stem Cell Research, National Research Council (2005) *Guidelines for Human Embryonic Stem Cell Research.* (National Academies Press, Washington, D.C.).

5. Vogel, G., and Holden, C. (2007) Developmental biology. Field leaps forward with new stem cell advances. *Science.* 318, 1224–1225.

6. Balanced Budget Downpayment Act. (1996) Public Law 104–99, Section 112.

7. National Institutes of Health (2007) *NIH Mission.* http://www.nih.gov/about/index.html#mission.

8. Greene, J. C. (1984) *American Science in the Age of Jefferson.* (Iowa State University Press, Ames, Iowa).

9. Mishan, E. J. (1971) On making the future safe for mankind. *J Public Interest.* 23, 33–61. Quoted in *New Atlantis*, Summer 2005, 96.

10. U.S. Environmental Protection Agency (2007) *Final Nanotechnology White Paper.* http://www.epa.gov/osa/nanotech.htm.

11. Domestic Policy Council, Office of Science and Technology Policy (2006) *American Competitiveness Initiative: Leading the World in Innovation.* http://www.whitehouse.gov/stateoftheunion/2006/aci/aci06-booklet.pdf.

12. National Institutes of Health (2003) *Investments, Progress, and Plans: Selected Examples from FY 1999–2003.* http://www.nih.gov/about/investments.htm.

13. National Institutes of Health (2003) *Fact Sheet: Doubling Accomplishments—Selected Examples.* http://www.nih.gov/about/researchresultsforthepublic/doublingaccomplishments.pdf.

14. Passell, P. (2000) *Exceptional Returns: The Economic Value of America's Investment in Medical Research.* (Funding First Executive Committee, Lasker Foundation, New York, N.Y.).

15. Aaron, H. J. (2002) The unsurprising surprise of renewed health care cost inflation. *Health Affairs Web Exclusive*, January 8. http://content.healthaffairs.org/cgi/reprint/hlthaff.w2.85v1.pdf.

16. Agnew, B. (1999) Varmus' last hurrah. *Washington FAX*, F-D-C Reports, December 6. (Archived by FDC reports, http://www.fdcreports.com).

17. Sarewitz, D. (1996) *Frontiers of Illusion*. (Temple University Press, Philadelphia, Penn.).

18. Bush, V. (1945) *Science—The Endless Frontier*. (U.S. Government Printing Office, Washington, D.C.).

19. U.S. Office of Management and Budget (2004) *Program Assessment: National Institutes of Health—Extramural Research Programs*. http://www.whitehouse.gov/omb/expectmore/summary/ 10002176.2004.html.

20. Committee on Science, Engineering, and Public Policy, National Academy of Sciences, National Academy of Engineering, Institute of Medicine (1999) *Evaluating Federal Research Programs: Research and the Government Performance and Results Act*. (National Academies Press, Washington, D.C.).

21. Specter, M. (2006) Political science: The Bush administration's war on the laboratory. *New Yorker*, March 13, pp. 58–67.

22. National Science Foundation (2006) Science and technology: Public attitudes and understanding. In *Science and Engineering Indicators 2006*. http://www.nsf.gov/statistics/seind06/c7/c7h.htm.

23. Center for Science in the Public Interest (2006) *FDA Panel Investigating Labeling of Antihypertensive Drugs Stacked with Industry Consultants*. http://www.cspinet.org/integrity/press/200604241.html.

24. Harris, G., and Berenson, A. (2005) 10 voters on panel backing pain pills had industry ties. *New York Times*, February 25.

25. Office of Technology Assessment (1991) *Federally Funded Research: Decisions for a Decade*. (U.S. Government Printing Office, Washington, D.C.).

26. National Science Foundation (2006) Academic research and development. In *Science and Engineering Indicators 2006*. http://www.nsf.gov/statistics/seind06/c5/c5i.htm.

27. National Science Foundation (2004) *NSF Grant Proposal Guide*. NSF 04–23. (National Science Foundation, Washington, D.C.).

28. American Association for the Advancement of Science (2007) *R&D Earmarks Headed Toward Records in 2007*. http://www.aaas.org/spp/ rd/earm07s.htm.

29. National Science Foundation. (2008) Appendix table 5–7: Federal obligations for academic research, by agency: 1970–2007. In *Science*

and *Engineering Indicators 2008*. http://www.nsf.gov/statistics/seind08/append/c5/at05-07.pdf.

30. National Institutes of Health (2007) *Research Project Success Rates*. http://report.nih.gov/award/success.cfm.

31. Marks, A. R. (2004) The economy of science. *J Clin Invest.* 114, 871.

32. President's Science Advisory Committee (1960) *Scientific Progress, the Universities, and the Federal Government*. (President's Science Advisory Committee, Washington, D.C.).

33. Liu, M., and Mallon, W. T. (2004) Tenure in transition: Trends in basic science faculty appointment policies at U.S. medical schools. *Acad Med.* 79, 205–213.

34. National Institutes of Health, Center for Scientific Review (2006) *How Scientists Are Selected for Study Section Service*. http://cms.csr.nih.gov/PeerReviewMeetings/StudySectionReviewers/HowScientistsareSelected+orStudySectionService.htm.

35. U.S. General Accounting Office (1994) *Peer Review: Reforms Needed to Ensure Fairness in Federal Agency Grant Selection*. (U.S. Government Printing Office, Washington, D.C.).

36. Merton, R. K. (1942) The normative structure of science. Reprinted in *The Sociology of Science: Theoretical and Empirical Investigations*. Ed. Merton, R. K., and Storer, N. W. (University of Chicago Press, Chicago, Ill., 1973), pp. 267–278.

37. U.S. House of Representatives, Committee on Science and Technology, Subcommittee on Investigations and Oversight (1981) *Fraud in Biomedical Research*. (U.S. Government Printing Office, Washington, D.C.).

38. U.S. Department of Health and Human Services (1989) *Responsibilities of PHS Awardee and Applicant Institutions for Dealing with and Reporting Possible Misconduct in Science: Final Rule*. 42 CFR 50, subpart A.

39. Schachman, H. K. (1990) *Congressional Testimony to the Subcommittee on Investigations on Oversight, Committee on Science, Space, and Technology, U.S. House of Representatives*. (U.S. Government Printing Office, Washington, D.C.).

40. Holton, G. (1973) *Thematic Origins of Scientific Thought: Kepler to Einstein*. (Harvard University Press, Cambridge, Mass.).

41. Jackson, C. I. (1984) *Honor in Science.* (Sigma Xi, The Scientific Research Society, New Haven, Conn.).
42. Goodstein, D. (2001) In the case of Robert Andrews Millikan. *Am Sci.* 89, 54–60.
43. Panel on Scientific Responsibility and the Conduct of Research, National Academy of Sciences, National Academy of Engineering, Institute of Medicine (1992) *Responsible Science: Vol. 1. Ensuring the Integrity of the Research Process.* (National Academies Press, Washington, D.C.).
44. Holmes, F. L. (2006) *Reconceiving the Gene: Seymour Benzer's Adventures in Phage Genetics.* (Yale University Press, New Haven, Conn.).
45. Office of Science and Technology Policy (2000) Federal policy on research misconduct. *Fed Reg.* 65, 76260–76264.
46. National Academies (2003) *Policy on Committee Composition and Balance and Conflicts of Interest for Committees Used in the Development of Reports.* http://www.nationalacademies.org/coi/bi-coi_form-0.pdf.
47. Hull, D. L. (1988) A mechanism and its metaphysics: An evolutionary account of the social and conceptual development of science. *Biol Philos.* 3, 123–155.
48. Hamilton, A. (1788) The Federalist No. 72. http://www.constitution.org/fed/federa72.htm.
49. Schachman, H. K. (2006) From "publish or perish" to "patent and prosper." *J Biol Chem.* 281, 6889–6903.
50. National Institutes of Health, U.S. Department of Health and Human Services (1999) Principles and guidelines for recipients of NIH research grants and contacts on obtaining and disseminating biomedical research resources: Final notice. *Fed Reg.* 64, 72090–72096.
51. Thomson, J. A., Itskovitz-Eldor, J., Shapiro, S. S., Waknitz, M. A., Swiergiel, J. J., et al. (1998) Embryonic stem cell lines derived from human blastocysts. *Science.* 282, 1145–1147.
52. Hwang, W. S., Ryu, Y. J., Park, J. H., Park, E. S., Lee, E. G., et al. (2004) Evidence of a pluripotent human embryonic stem cell line derived from a cloned blastocyst. *Science.* 303, 1669–1674.

53. University of Pittsburgh Investigative Board (2006) *Summary Investigative Report on Allegations of Possible Scientific Misconduct on the Part of Gerald P. Schatten, Ph.D.* http://www.physics.utah. edu/~detar/phys4910/readings/misconduct/Gerald_Schatten_Final_ Report_2.08.pdf.
54. Franklin, B. (1868) *The Autobiography of Benjamin Franklin.* (Reprint of the 1868 John Bigelow edition, Dover Publications, Mineola, N.Y., 1996).
55. Blumenstyk, G. (2005) Colleges cash in on commercial activity. *Chron Higher Educ.* 52, A25.
56. *Madey v. Duke University.* 307 F.3d 1351, 1362 (Fed. Cir. 2002).
57. U.S. House of Representatives, Committee on Government Operations (1990) *Are Scientific Misconduct and Conflicts of Interest Hazardous to Our Health?* (U.S. Government Printing Office, Washington, D.C.).
58. U.S. Public Health Service (1995) Objectivity in research. NOT-95-179. *NIH Guide.* 24(25). http://grants.nih.gov/grants/ guide/notice-files/not95-179.html.
59. McCrary, S. V., Anderson, C. B., Jakovljevic, J., Khan, T., McCullough, L. B., et al. (2000) A national survey of policies on disclosure of conflicts of interest in biomedical research. *N Engl J Med.* 343, 1621–1626.
60. Task Force on Financial Conflicts of Interest in Clinical Research (2002) *Protecting Subjects, Preserving Trust, Promoting Progress II: Principles and Recommendations for Oversight of an Institution's Financial Interests in Human Subjects Research.* http://www.aamc.org/ research/coi/2002coireport.pdf.
61. U.S. General Accounting Office (2001) *Biomedical Research: HHS Direction Needed to Address Financial Conflicts of Interest.* (U.S. Government Printing Office, Washington, D.C.).
62. Krimsky, S. (2005) The funding effect in science and its implications for the judiciary. *J Law Policy.* 13, 43–68.
63. Angell, M. (2000) Is academic medicine for sale? *N Engl J Med.* 342, 1516–1518.
64. National Institutes of Health (2005) *Summary of NIH-Specific Amendments to Conflict of Interest Ethics Regulations.* http://www.nih. gov/about/ethics/summary_amendments_08252005.htm.

Chapter 5: Informed Consent and Risk

1. Lemonick, M. D., and Goldstein, A. (2002) At your own risk. *Time*, April 22.
2. Rothman, D. J. (1991) *Strangers at the Bedside*. (Basic Books, New York, N.Y.).
3. Beecher, H. K. (1966) Ethics and clinical research. *N Engl J Med.* 274, 1354–1360.
4. Brunner, B. (2004) *The Tuskegee Syphilis Experiment*. http://www.tuskegee.edu/Global/Story.asp?s=1207586.
5. U.S. Department of Health, Education, and Welfare, Office of the Secretary (1979) *The Belmont Report*. http://www.hhs.gov/ohrp/humansubjects/guidance/belmont.htm.
6. Solomon, W. D. (1995) Normative ethics theories. In *Encyclopedia of Bioethics: Vol. 2*. Ed. Reich, W. T. (Simon & Schuster Macmillan, New York, N.Y.), pp. 736–747.
7. Levine, R. J. (1986) *Ethics and Regulation of Clinical Research*, 2nd ed. (Yale University Press, New Haven, Conn.).
8. Crawford, L. M. (2004) *Speech before Global Pharmaceutical Strategies Seminar: Remarks*. http://www.fda.gov/oc/speeches/2004/gpss0525.html.
9. European Medicines Agency (2001) *Choice of Control Group in Clinical Trials*. ICH Topic E10. http://www.emea.europa.eu/pdfs/human/ich/036496en.pdf.
10. Angell, M. (1997) The ethics of clinical research in the third world. *N Engl J Med.* 337, 847–849.
11. Grinnell, F. (1990) Endings of clinical research protocols: Distinguishing therapy from research. *IRB Ethics Hum Res.* 12, 1–3.
12. Faden, R. R., and Beauchamp, T. L. (1986) *A History and Theory of Informed Consent*. (Oxford University Press, New York, N.Y.).
13. Appelbaum, P. S., Roth, L. H., Lidz, C. W., Benson, P., and Winslade, W. (1987) False hopes and best data: Consent to research and the therapeutic misconception. *Hastings Cent Rep.* 17, 20–24.
14. Natanson, M. (1986) *Anonymity: A Study in the Philosophy of Alfred Schutz*. (Indiana University Press, Bloomington, Ind.).
15. Institute of Medicine (2001) *Preserving Public Trust: Accreditation and Human Research Participant Protection Programs*. (National Academies Press, Washington, D.C.).

16. Shaw, S., and Barrett, G. (2006) Research governance: Regulating risk and reducing harm? *J R Soc Med.* 99, 14–19.

17. Editorial Board (1971) The body: From baby hatcheries to "xeroxing" human beings. *Time,* April 19.

18. Pearson, H. (2006) What is a gene? *Nature.* 441, 399–401.

19. Bickel, K. S., and Morris, D. R. (2006) Silencing the transcriptome's dark matter: Mechanisms for suppressing translation of intergenic transcripts. *Mol Cell.* 22, 309–316.

20. Levy-Lahad, E., and Plon, S. E. (2003) Cancer. A risky business—assessing breast cancer risk. *Science.* 302, 574–575.

21. Osteogenesis Imperfecta Foundation (2007) *Home Page.* http://www.oif.org/site/PageServer.

22. March of Dimes (2007) *Down Syndrome.* http://www.marchofdimes.com/professionals/14332_1214.asp.

23. Fitzgerald, F. T. (n.d.) *Doctors and Delphi.* Unpublished manuscript.

24. Allen, G. E. (1986) The Eugenics Record Office at Cold Spring Harbor, 1910–1940: An essay in institutional history. *Osiris.* 2, 225–264.

25. *Buck v. Bell,* 274 U.S. 200 (1927).

26. UT Southwestern Medical Center Institutional Review Board (2007) *DNA Consent.* http://www8.utsouthwestern.edu/utsw/cda/dept27777/files/55431.html.

27. Egozcue, J., Santalo, J., Gimenez, C., Perez, N., and Vidal, F. (2000) Preimplantation genetic diagnosis. *Mol Cell Endocrinol.* 166, 21–25.

28. Hardy, K., Spanos, S., Becker, D., Iannelli, P., Winston, R. M., et al. (2001) From cell death to embryo arrest: Mathematical models of human preimplantation embryo development. *Proc Natl Acad Sci USA.* 98, 1655–1660.

29. Doerflinger, R. M. (1999) The ethics of funding embryonic stem cell research: A Catholic viewpoint. *Kennedy Inst Ethics J.* 9, 137–150.

30. ASHG Social Issues Subcommittee on Familial Disclosure (1998) ASHG statement. Professional disclosure of familial genetic information. The American Society of Human Genetics Social Issues Subcommittee on Familial Disclosure. *Am J Hum Genet.* 62, 474–483.

31. U.S. Department of Health and Human Services (2003) *Summary of the HIPAA Privacy Rule.* http://www.hhs.gov/ocr/privacysummary.pdf.

32. American Association for Public Opinion Research (2006–2007) *IRB FAQs for Survey Researchers*. http://www.aapor.org/irbfaqsforsurveyresearchers.

33. U.S. House of Representatives Committee on Government Report (2000) *Human Subject Research Protections*. (U.S. Government Printing Office, Washington, D.C.).

34. McCabe, E. R. B. (2003) *Legacy of the Secretary's Advisory Committee on Genetic Testing*. http://www4.od.nih.gov/oba/sacghs/meetings/June2003/Presentations/McCabe_t.pdf.

35. Wadman, M. (2000) Geneticists oppose consent ruling. *Nature*. 404, 114–115.

36. American Society of Human Genetics (2000) *Family History and Privacy Advisory—March, 2000*. http://www.ashg.org/pages/statement_32000.shtml.

37. Indigenous Peoples Council on Biocolonialism (2006) *Genographic Project Put on Hold in North America*. http://www.ipcb.org/issues/human_genetics/htmls/update_1206.html.

38. American Board of Medical Specialties (2007) *Medical Genetics*. http://www.abms.org/Who_We_Help/Consumers/About_Physician_Specialties/medical.aspx

39. Paul, D. B. (1997) The history of newborn phenylketonuria screening in the United States. In *Promoting Safe and Effective Genetic Testing in the United States: Final Report of the Task Force on Genetic Testing*. Ed. Holtzman, N. A., and Watson, M. S. (National Human Genome Research Institute, National Institutes of Health, Bethesda, Md.). http://www.genome.gov/10002397.

40. National Institute of Neurological Disorders and Stroke (2007) *NINDS Tay-Sachs Disease Information Page*. http://www.ninds.nih.gov/disorders/taysachs/taysachs.htm.

41. Journal of Gene Medicine (2008) *Gene Therapy Clinical Trials Worldwide*. http://www.wiley.co.uk/wileychi/genmed/clinical/.

42. Jonas, H. (1974) Biological engineering—a preview. In *Philosophical Essays: From Ancient Creed to Technological Man*. (University of Chicago Press, Chicago, Ill.), pp. 141–167.

43. Office of Technology Assessment (1984) *Human Gene Therapy: Background Paper*. (U.S. Government Printing Office, Washington, D.C.).

44. Couzin, J., and Kaiser, J. (2005) Gene therapy. As Gelsinger case ends, gene therapy suffers another blow. *Science.* 307, 1028.

45. Office of Technology Assessment (1988) *Mapping Our Genes— Genome Projects: How Big? How Fast?* (U.S. Government Printing Office, Washington, D.C.).

46. Institute of Medicine (2002) *Responsible Research: A Systems Approach to Protecting Research Participants.* (National Academies Press, Washington, D.C.).

47. Stolberg, S. G. (1999) *F.D.A. Officials Fault Penn Team in Gene Therapy Death.* http://query.nytimes.com/gst/fullpage.html?res= 9F0CE0DA1631F93AA35751C1A96F958260&sec=health.

48. Kipnis, K. (2001) Vulnerability in research subjects: A bioethical taxonomy. In *Ethical and Policy Issues in Research Involving Human Subjects: Vol. 2. Commissioned Papers.* (National Bioethics Advisory Commission, Washington, D.C.), pp. G1–G13.

49. Grinnell, F. (2004) Subject vulnerability: The precautionary principle of human research. *Am J Bioethics.* 4, 72–74.

50. Institute of Medicine (1998) *Scientific Opportunities and Public Needs: Improving Priority Setting and Public Input at the National Institutes of Health.* (National Academies Press, Washington, D.C.).

51. U.S. Code of Federal Regulations (2005) Title 45, Public Welfare. Part 46, Protection of Human Subjects. Section 46.107, IRB Membership. 45 *CFR* 46.107.

52. National Bioethics Advisory Committee (2001) *Ethical and Policy Issues in Research Involving Human Participants.* (National Bioethics Advisory Commission, Washington, D.C.).

53. U.S. Department of Health and Human Services, Office for Protection from Research Risks (1996) *"Exculpatory Language" in Informed Consent.* http://www.hhs.gov/ohrp/humansubjects/guidance/exculp.htm.

54. Steinbrook, R. (2006) Compensation for injured research subjects. *N Engl J Med.* 354, 1871–1873.

Chapter 6: Faith

1. Nelkin, D. (2004) God talk: Confusion between science and religion (posthumous essay). *Sci Technol Hum Values.* 29, 139–152.

2. U.S. District Court (2005) *Tammy Kitzmiller, et al. vs. Dover Area School District.* 400 F.Supp.2d 707 (M.D.Pa.). http://www.pamd. uscourts.gov/kitzmiller/kitzmiller_342.pdf.

3. Wertz, D. C. (2002) Embryo and stem cell research in the United States: History and politics. *Gene Ther.* 9, 674–678.

4. Singer, P. (1993) *Practical Ethics.* (Cambridge University Press, Cambridge, U.K.).

5. Ericson, E. L. (1988) *The Humanist Way: An Introduction to Ethical Humanist Religion.* http://ethicalunion.org/ericson2.html.

6. Barbour, I. G. (1990) *Religion in an Age of Science.* (Harper, San Francisco, Calif.).

7. Bohr, N. (1929) Introductory survey. Reprinted in *Philosophical Writings of Niels Bohr: Vol. 1.* (Ox Bow Press, Woodbridge, Conn., 1987), pp. 1–24.

8. Pope John Paul II (1980) Address at the Einstein session of the Pontifical Academy of Science, 10 November 1979. *Science.* 207, 1165–1167.

9. Lockwood, R. P. (2000) *Galileo and the Catholic Church.* http://www. catholicleague.org/research/galileo.html.

10. Dickinson, E. (1864) A Light Exists in Spring. Reprinted in *The Complete Poems of Emily Dickinson.* (Back Bay Books, Boston, Mass., 1976), p. 395.

11. James, W. T. (1961) *The Varieties of Religious Experience: A Study in Human Nature.* (Collier Macmillan, New York, N.Y.).

12. Einstein, A. (1932) *My Credo.* http://www.einsteinandreligion.com/ credo.html.

13. Strauss, L. (1997) Jerusalem and Athens: Some preliminary reflections. In *Jewish Philosophy and the Crisis of Modernity.* Ed. Strauss, L., and Green, H. G. (State University of New York Press, New York, N.Y.), pp. 377–408.

14. Cohen, H. (1993) The social ideal as seen by Plato and by the Prophets. In *Reason and Hope.* Ed. Jospe, E. (Wayne State University Press, Detroit, Mich.), pp. 66–77.

15. Soloveitchik, J. B. (1992) *The Lonely Man of Faith.* (Doubleday, New York, N.Y.).

16. (2007) *The Mission of the Franklin Institute.* http://www2.fi.edu/ join/giving/individual/dev_mission.php.

17. Hume, D. (1748) *An Inquiry Concerning Human Understanding.* (Reprint: Bobbs-Merrill, Indianapolis, Ind., 1955).

18. Whitehead, A. N. (1925) *Science and the Modern World.* (Free Press, New York, N.Y.).

19. Einstein, A. (1954) Science and religion. In *Idea and Opinions.* (Crown, New York, N.Y.), pp. 41–49.

20. James, W. (1907) Pragmatism's conception of truth. Reprinted in *Pragmatism and the Meaning of Truth.* (Harvard University Press, Cambridge, Mass., 1975), pp. 95–113.

21. Soloveitchik, J. B. (1983) *Halakhic Man.* (Jewish Publication Society, New York, N.Y.).

22. Whitman, W. (1855) *Leaves of Grass.* http://www.gutenberg.org/etext/1322.

23. Buber, M. (1923) *I and Thou.* (Reprint: Charles Scribner's Sons, New York, N.Y., 1970).

24. Friedman, M. (1981) *Martin Buber's Life and Work: The Early Years.* (E. P. Dutton, New York, N.Y.).

25. Otto, R. (1932) *Mysticism East and West.* (Macmillan, New York, N.Y.).

26. European Network of Scientific Co-operation on Medicine and Human Rights (1998) *The Human Rights, Ethical and Moral Dimensions of Health Care—120 Practical Case Studies.* (Council of Europe Publishing, Strasbourg, France).

27. Curran, C. E. (2007) Personal communication with the author, 8 November.

28. Maimonides, M. (1956) *The Guide for the Perplexed.* (Dover, New York, N.Y.).

29. The Council of Trent (1546) *Decree Concerning the Edition, and the Use, of the Sacred Books.* http://history.hanover.edu/texts/trent/trentwh.html.

30. Plaut, M. (2005) Why We Censor. In *Dei'ah ve Dibur,* December 21. http://chareidi.shemayisrael.com/archives5766/vayeishev/ochankvyv66.htm.

31. State of Texas Republican Party (2006) *2006 State Republican Party Platform.* http://www.texasgop.org/site/DocServer/Platform_Updated.pdf?docID=2001.

32. National Center for Science Education (2007) *About NCSE.* http://www.ncseweb.org/about.asp.

33. Miller, J. D., Scott, E. C., and Okamoto, S. (2006) Science communication. Public acceptance of evolution. *Science.* 313, 765–766.

34. Johnson, P. E. (2007) *Intelligent Design in Biology: The Current Situation and Future Prospects.* Center for Science and Culture, February 19. http://www.discovery.org/scripts/viewDB/ index.php?command=view&id=3914&program=CSC%20-% 20Scientific%20Research%20and%20Scholarship%20-% 20History%20and%20Philosophy%20of%20Science.

35. Behe, M. J. (1994) Experimental support for regarding functional classes of proteins to be highly isolated from each other. In *Darwinism: Science or Philosophy.* Ed. Buelland, J., and Hearn, V. (Foundation for Thought and Ethics, Richardson, Tex.), pp. 60–67.

36. Moore, J. A. (1972) *Heredity and Development.* (Oxford University Press, New York, N.Y.).

37. Gilbert, S. F. (2006) *Developmental Biology.* (Sinauer, New York, N.Y.).

38. Einstein, A. (1954) Religion and science. In *Idea and Opinions.* (Crown, New York, N.Y.), pp. 36–40.

39. Ruse, M. (2005) *The Evolution-Creation Struggle.* (Harvard University Press, Cambridge, Mass.).

40. Steering Committee on Science and Creationism (1999) *Science and Creationism: A View from the National Academy of Sciences.* (National Academy Press, Washington, D.C.).

41. Pollack, R. (2000) *Faith of Biology and the Biology of Faith.* (Columbia University Press, New York, N.Y.).

42. Bohr, N. (1928) The quantum postulate and the recent development of atomic theory. *Nature.* 128 (suppl.), 580–590.

43. Stone, C. D. (1987) *Earth and Other Ethics: The Case for Moral Pluralism.* (Harper & Row, New York, N.Y.).

44. James, W. (1896) The will to believe. Reprinted in *Selected Writings.* (J. M. Dent, London, U.K., 1997), pp. 249–270.

45. Holton, G. (1988) *Thematic Origins of Scientific Thought.* (Harvard University Press, Cambridge, Mass.).

46. Bohr, N. (1929) The atomic theory and the fundamental principles underlying the description of nature. Reprinted in *The Philosophical Writings of Niels Bohr: Vol. 1.* (Ox Bow Press, Woodbridge, Conn., 1987), pp. 102–119.

47. Honner, J. (1982) Niels Bohr and the mysticism of nature. *Zygon.* 17, 243–253.

48. Bohr, N. (1960) Unity of human knowledge. Reprinted in *Philosophical Writings of Niels Bohr: Vol. 3.* (Ox Bow Press, Woodbridge, Conn., 1987), pp. 8–16.

49. Alexander, P. (1956) Complementary Descriptions. *Mind N.S.* 65, 145–165.

50. Burhoe, R. W. (1954) Religion in the age of science. *Science.* 120, 522–524.

51. Grinnell, F. (1994) Radical intersubjectivity: Why naturalism is an assumption necessary for doing science. In *Darwinism: Science or Philosophy.* Ed. Buelland, J., and Hearn, V. (Foundation for Thought and Ethics, Richardson, Tex.), pp. 99–106.

52. American Association for the Advancement of Science (2007) *AAAS Dialogue on Science, Ethics and Religion: Mission Statement.* http://www.aaas.org/spp/dser/01_About/01_index.shtml.

53. Babylonian Talmud (1948) *The Soncino Talmud, English Translation.* (Oxford University Press, Oxford, U.K.).

54. Dukas, H., & Hoffman, B. (1981) *Albert Einstein the Human Side.* (Princeton University Press, Princeton, N.J.).

Index

Page numbers in *italics* indicate figures or tables.